Television in the
Corporate Interest

Richard Bunce

The Praeger Special Studies program—utilizing the most modern and efficient book production techniques and a selective worldwide distribution network—makes available to the academic, government, and business communities significant, timely research in U.S. and international economic, social, and political development.

Television in the Corporate Interest

PRAEGER SPECIAL STUDIES IN U.S. ECONOMIC, SOCIAL, AND POLITICAL ISSUES

Praeger Publishers   New York   Washington   London

Library of Congress Cataloging in Publication Data

Bunce, Richard.
    Television in the corporate interest.

    (Praeger special studies in U.S. economic, social, and
political issues)
    Includes bibliographical references and index.
    1. Television broadcasting—United States.
I. Title.
HE8700.8.B85   384.55'0973   75-23958
ISBN 0-275-55950-5

PRAEGER PUBLISHERS
111 Fourth Avenue, New York, N.Y. 10003, U.S.A.

Published in the United States of America in 1976
by Praeger Publishers, Inc.

Printed in the United States of America

CONTENTS

LIST OF TABLES

1

INTRODUCTION: TOWARDS A COMMUNICATIONS DEMOCRACY

"Nearly forty years of the present broadcasting system has conditioned a generation of citizens to find it normal. Most people find it
difficult to think of other kinds of broadcast systems and uses."[1] This
was written a decade ago by the then president of the National Association of Educational Broadcasters, Harry Skornia. These are encouraging words to those who built, operate and now control the American
broadcast communications system—encouraging, but not surprising.
For they are echoed by Skornia's commercial counterpart, Vincent
Wasilewski, president of the National Association of Broadcasters.
Wasilewski declared a few years later that "broadcasting stands as
the most successful and universally accepted business enterprise in
history."[2]

It would be mistaken to dismiss Skornia's claim as only a veiled
pleading for monetary support for educational broadcasting; or
Wasilewski's, as simply the public relations prattle of the commercial
broadcasters' chief lobbyist. For it is evidence of their successful
grip on our conception of social possibilities that ABC (American
Broadcasting Company), CBS (Columbia Broadcasting System), and
NBC (National Broadcasting Company) have become generic terms for
the television technology itself.

It is not so much that alternative broadcast systems and uses
have failed to be imagined; rather, that such visions have been effec-
tively isolated from public consciousness. Two factors in particular
seem to encourage this.

The first concerns the rapid technological changes which have
proceeded since the advent of mass electronic communication. The
development and introduction first of radio, followed by frequency
modulation, television, cable and satellite technologies have all occurred in the space of forty-five years, roughly the period 1920-65.
In the early period the radio equipment manufacturers settled upon a

1

use for their machinery which promised the most monetary return in terms of mass receiver set sales. Organizational structures appropriate for these goals were established under their authority. And these same control structures were reproduced as each new technological invention was introduced by the manufacturing wings of these enterprises. As an indication of the rapidity of these technical changes and the comparatively short time span involved, we see that even the same personnel (William Paley and Frank Stanton of CBS, General David Sarnoff of RCA), who in the 1920s designed the social use of radio, were in 1970 still presiding over a now much more dense and differentiated technological landscape.

The rapidity of innovation has worked to their advantage. Electronics, from the wireless to the computer and the satellite, has redeemed the belief in progress with a belief in technology. Technological innovation is the corporate state's equivalent to Adam Smith's invisible hand. Through a conception of the world unfolding in accordance with laisser-innover, we are charmed into believing that technological innovation, given a free hand, is the ultimate problem solver. This belief in emancipation through technological hardware is contradicted by historical developments, but it persists. Forecasts of technological emancipation have accompanied each electronic innovation beginning with radio. Today expectations are raised by promises that cable and video-tape technology will provide for the information diversity, the feedback capacity and other necessities that, unfortunately, older technologies simply could not facilitate.

In fact, these are old promises and pre-existing potentials. They were spoken when amplitude modulation, frequency modulation, and television were each in their respective incubation periods. The fact that such expectations have not been realized with each successive innovation should provoke skepticism at the very least with regard to the packaged promises attributed to the latest products from the manufacturers' laboratories. But the rapid and accelerating reproduction of novel hardware has captured imagination and compressed the time available for experimenting and developing alternative use systems. Consequently industrial economic structures have been superimposed and reproduced on developing information structures and we are left with Skornia's observation that "most people find it difficult to think of other kinds of broadcast systems and uses."

A second factor which shields from public consciousness the possibilities for alternative communications arrangements is the political culture which has informed the development of mass communications technologies. The term "political culture" is necessarily broad. It is used here to refer to the fact that electronic communications were first developed and deployed in the context of military institutions during the First World War. And ever since then they have capabilities to mass audiences on the order of present-day commercial networks, but rather to permit electronic feedback of the computer variety, useful for marketing and purchasing transactions, as well as private video communication, on the order of telephony.

The potential for mass communication in the sense of mass participation in the communication process, under present control structures, will likely remain some future promise forever. But its potential has been with us since the beginning of broadcasting, and with the earliest technologies. The capacity for feedback and the opportunity for people to take the transmitter role, and not merely the audience-reception role, are not precluded by the prevailing technology of broadcasting. As German communications scholar H. M. Enzensberger has explained, the prevailing media control system

> . . . allows no reciprocal action between transmitter and receiver; technologically speaking it reduces feedback to the lowest point compatible with the system.
>
> This state of affairs however cannot be justified technically. On the contrary. Electronic techniques recognize no contradiction in principle between transmitter and receiver. Every transistor radio is, by the nature of its construction, at the same time a potential transmitter; it can interact with other receivers by circuit reversal. The development from a mere distribution medium to a communications medium is technically not a problem. It is consciously prevented for understandable political reasons. The technical distinction between receivers and transmitters reflects the social division of labor into producers and consumers, which in the consciousness industry becomes of particular political importance.[6]

In other words there is nothing in the electronic media to fix inherently the roles of communicator (producer) and audience (consumer). Rather the roles, and their attendant power inequalities, must be enforced through economic, legal and administrative measures.

Whether fashioning alternative visions, or plumbing temporal social arrangements for their essences, ultimately one must confront the nature of power inequalities. This is as true of the power to communicate as it is of more familiar power hierarchies such as economic and political inequalities. Of course, this is exactly what Brecht's proposal forces us to do. Juxtaposing the idea of mass participation in electronic communications with the reality of centrally controlled electronic media exposes the inequalities inherent in an essentially distributive system.

In the centrally managed communications system, the distinctions between producers of information and consumers are a function of power. An egalitarian system transforms these distinctions, making them only a function of time. With audiences freely able to produce, that is, feed into or feedback, as well as consume, the roles are no longer fixed, but fluid. Only when they become rigid is it meaningful to speak of communicator-audience role differences at all. In a system that is fluid over time, these distinctions are no more salient or interesting than differences in initiation and response in private conversation.

Beyond the idea of communications reciprocity in electronic media, Brecht redefines the information architecture itself with the idea of audiences organized for the production of messages. This is critical for here is an assault upon the concept of "audience" that is created by the separation of people from the technology of transmission. One must understand that in a system which universally distributes receiver sets and strictly controls the availability of and access to transmitters, an audience is created which is not only characterized by passivity, in the consumer sense, but is also atomized. In such a system, the flow of information atomizes the audience: the audience is created as a collectivity of isolated individuals, prevented from realizing their relatedness, prevented from becoming a group. The message flow in this system becomes dissociative communication. This This is self-contradictory, because it has no basis in community, which is the raison d'etre of communicating. Those in control of the transmitters have access to the audience, but the audience has neither access to itself, nor to those in control. What is called communication in this system functions to break up group linkages, to dissociate those plugged into the technology, to atomize participation.

While this may strike some as abstract, it is actually fundamental to our trained incapacity to consider truly alternative modes of electronic communications, which we spoke of earlier. Since our experience with electronic mass media has uniformly been as audience, we are accustomed, albeit unconsciously, to imagine ourselves as atomized. Accordingly, the notion of mass participation in electronic communications must immediately suggest the specter of millions of individuals independently seeking access. While this is congruent with our experience as audience, it is also part of the superstructure of the existing system of media control, which proceeds, characteristically, to equate phone-in programs with genuine participation and meaningful access.

Hence the notion of organizations of audiences for massive participation in electronic communications is revolutionary both for the nature of information structures and for the meaning of communications in an electronic age.

There are no concrete examples of the kind of egalitarian information structure envisioned here. But a suggestive approximation to communications democracy was briefly in force in the Netherlands from 1967-72, and it is worth considering the outlines of that system.[7]

Historically Dutch broadcasting is an exceptional case because the system is managed and operated neither by the state nor by private commercial corporations. Instead a number of very large audience-membership associations were formed reflecting the predominant religious, political, and economic cleavages in Dutch society. These organizations, established to provide mass communications reflecting their particular interests, are state licensed and share equally the transmission facilities and broadcast time on the Netherlands' two television channels. From 1967-72 this system was expanded, as access was opened to virtually any group seeking it. Any audience association with 15,000 or more members was licensed for regular weekly broadcasts. And 10 percent of the total broadcast time was reserved for other political, religious, and cultural movements (not audience organizations per se), which sought to propagate their views over the air on an occasional basis. No membership requirements were made of these latter organizations.

While the Dutch experiment could not be described as a fully accessible, two-way mass electronic medium, it nonetheless provides a qualitatively different model from that of the centrally managed information flow. The plural audience associations which generate tele-communications have a membership constituency, and thus an accountability, absent in state and commercially controlled systems. The public access provisions, while certainly limited, still introduce the possibility of feedback and reciprocal (or two-way) communications, even for those social movements which are numerically marginal. But it is the characteristic of the Dutch system of audiences organized for mass communication of their own interests that is most engaging, innovative, and also democratic.

Contrast the Brechtian vision, or the fledgling experiments of the Dutch, with an authoritarian information structure and communications flow—one that is centrally controlled and managed by a coalition of elite groups which select and produce the information and images for the whole of society. The direction of the flow itself suggests that the overwhelming majority in the society has no ideas or images worth communicating. Not only is this mistaken; it distorts the natural processes of cultural, social and political experimentation, innovation and diffusion by which egalitarian societies change, grow, develop, and survive. An authoritarian flow turns from these natural processes, creating a coalition of elite interests to filter, select and diffuse primarily those existing and emergent images and messages which conform to their special and exclusive needs.

The result is a scenario which was familiar, albeit briefly, to American television audiences in the late 1960s and early 1970s. Beginning with the peace movement, informationally disenfranchised and feedback-impoverished interest groups created dramatic events to secure exposure in the national communications channels, then they retreated to their TV screens to await some evidence of their existence. Only the resulting coverage, which they were to learn is not the same as exposure, rarely if ever verified and validated their experience. This kind of shock media therapy is quickly dissipated, moreover, when the novelty of such communiques recedes. Only when feedback is structured into the communications flow will it be enduring. Still, the dramatization of communications inequalities must be an aid in the propagation of alternative visions and systems even if it cannot transform existing ones.

To refine our understanding of communications inequalities in a centrally managed information structure, one in which access is an executive privilege, not a popular right, two factors must be considered. Of first importance is the system's willingness to present diverse and antagonistic interests, needs and priorities in its information flow. In other words, does the system provide opportunities for public forums, for the communication of social controversies and antagonisms sufficient to ameliorate significantly the authoritarian character of the information flow?

The second factor concerns the system's inherent capacity to itself represent diverse and antagonistic interests, needs and priorities. Applying this measure to the Dutch system, for example, raises this question: If only large audience associations are licensed for access, do these organizations as a group contain a sufficient social, economic and cultural pluralism of interests to provide communications of a truly diverse and antagonistic nature? The whole matter of a system's capacity to represent antagonisms becomes critical the more that each and every interest group cannot present itself through the system (that is, access is privileged), and the more that the system is unwilling to create a forum for interests other than its own.

Communications inequalities should be assessed jointly as a function of the diversity of interests that comprise the system, and as a function of the provision of access to interests not part of the system.

The question to be addressed in the pages that follow is basically this: Are the business corporations which at present control American radio and television either willing to present, or capable of representing, the diverse and antagonistic interests that are essential to the free flow of ideas and a democratic society? Are these business corporations responsible trustees of our social communications system? Many astute and informed critics of the system believe that they are. The failures, these liberal critics contend, stem not from business trusteeship per se, but rather from regulatory failures to halt monopolistic trends in the communications industry. Because this liberal viewpoint has been so important in shaping regulatory policy in communications for four decades; because it has become the orthodox diagnosis for nearly all serious critics of the system's ills; and because the continued legitimacy of this posture inhibits alternative visions and policies—for these reasons the liberal theories of business ownership and communications regulation deserve the test of evidence. That evidence will be presented and examined in the chapters that follow.

NOTES

1. Harry J. Skornia, Television and Society (New York: McGraw-Hill, 1965), p. 9.

2. Variety, January 7, 1970

3. See particularly Herbert Schiller, Mass Communications and American Empire (New York: Augustus Kelley, 1969); Erik Barnouw, The Image Empire (New York: Oxford University Press, 1970).

4. Fred S. Siebert, Theodore Peterson and Wilbur Schramm, Four Theories of the Press (Urbana: University of Illinois Press, 1956).

5. John Willett, editor and translator, Brecht on Theatre (New York: Hill and Wang, 1964), p. 52.

6. Hans Magnus Enzensberger, "Constituents of a Theory of the Media," New Left Review 64 (November-December 1970): 15.

7. Unfortunately, few accounts of the unique aspects of Dutch broadcasting have appeared in the English language press. The sources available concentrate on the legal structure, virtually neglecting the nature of the communications experience: J. van Santbrink, "Legislation and the Broadcasting Institutions in the Netherlands," Parts I, II, and III in EBU Review 98B (July 1966): 55-58; 102B (March 1967): 53-59; 116B (July 1969): 46-50. See also "Disappearance of the Broadcasting 'mini-groups,'" EBU Review 23, no. 6 (November 1972): 46-47; Burton Paulu, Radio and Television Broadcasting on the European Continent (Minneapolis: University of Minnesota Press, 1967), pp. 71-75; and Walter B. Emery, National and International Systems of Broadcasting: Their History, Operation and Control (East Lansing: Michigan State University Press, 1969), pp. 140-57.

2

BUSINESS PLURALISM

The question of who owns and controls a service like broadcast communications would seem to be self-evidently important. But the whole matter of television station ownership might never have become as critical as it now is if Congress had designated broadcasting a common carrier service. As a common carrier, broadcasting, like the telephone utility and some transportation services, would have been made available equally to all. Not only would provisions for reception be egalitarian, so too would the privilege of transmission, the opportunity to communicate with the public via the airwaves.

But Congress, in passing the Communications Act of 1934, did not choose this course. Instead it specifically designated broadcasting as a noncommon carrier service.[1] Consequently a license to broadcast carries with it the privilege of deciding what will be communicated, how, when, and by whom.

The free exercise of this privilege was likewise secured in this same law. Section 326 of the Act defines broadcasting as speech protected by the First Amendment. This means that the licensing authority, the Federal Communications Commission, is prevented by statute and the Constitution from censoring broadcast communications.

These legal requirements vest the television licensee with full authority over the use of his frequency. The licensee is legally empowered to decide who shall have access to the medium, and for what purpose. And there has been little in the way of Commission policy to alter this privilege.

Rather, under Federal Communications Commission (FCC) administration the privilege has become a property right, and licensees have become owners of frequencies which nominally belong to the public.

While the idea that the people own the airwaves was fundamental to the Communications Act of 1934, it survives today mostly as legal fiction. "Hooey and nonsense" was the term, in fact, which the President of the National Association of Broadcasters applied to this idea back in 1946.[2] FCC practices suggest a similar, if less candid, contempt for the notion. Theoretically, licenses are granted only for a three-year period, with renewal conditional upon the broadcaster's exercise of this public trust. The standard to be applied in such cases is the "public interest, convenience and necessity," to quote the Communications Act. But in spite of scores of license challenges by audience organizations, the Commission has never once denied or revoked a commercial television license for inadequate programming. In some thirty years of television licensing involving thousands of cases, the Commission has only once denied license renewal at the behest of a citizens group to a noncommercial broadcaster on programming grounds. That historic decision was taken in 1975 when the Commission refused renewal to the Alabama Educational Television Commission's eight licenses because of AETC's discrimination against blacks in program service.[3] So while the license renewal procedure serves in theory to check the exercise of the broadcast privilege, in fact the privilege, once vested, is never withdrawn. In effect this makes the frequency the property of the licensee, not the public.

If these arrangements create enduring communications inequalities, they also create inequalities with respect to free expression. Such reasoning has prompted Jerome Barron to write that "a realistic view of the First Amendment requires recognition that a right of expression is somewhat thin if it can be exercised only at the sufferance of the managers of mass communications."[4]

The legal and administrative framework, then, transfers authoritative control over broadcasting to private interests which are guaranteed the free exercise of their authority. Accordingly, no public right to media access is provided for. Under this system it is not the license to broadcast which is a privilege, but public access. Whether this authority is used narrowly to perpetuate inequalities, or contrarily, to facilitate egalitarian use in the interests of freer expression, becomes a prerogative of the owners. Whether this authority is used to exploit the owners' exclusive interests in communications, or contrarily, to provide "information from diverse and antagonistic sources" (which the Supreme Court declared essential to democratic communications), is likewise the owners' prerogative.[5] This, then, is why the question of communications ownership is so critical—because free expression over the public airwaves has been made to rest upon the sufferance of private owners.

This fact has not been lost on that body most responsible for it, the FCC. The Commission, like the Congress, early turned back public agitation for a social pluralism of licensees. Legislation requiring the Commission to allot one fourth of all frequencies to "educational, religious, agricultural, labor, cooperative, and similar nonprofit associations" was defeated in Congress in 1934, and the disenfranchised fared no better in subsequent Commission licensing decisions.[6]

Instead the FCC promised a business pluralism. For a time this meant an almost singularly commercial system. Later, educational channels were formally reserved, but only after twenty years of verbal skirmishes had made the religious, agricultural, labor, cooperative and other nonprofit interest groups yield hope of ever securing their own frequencies. The more stable and foundation-backed nonprofit contenders, the academic institutions, were ultimately rewarded for their persistence in the struggle with a fixed percentage of television frequencies.

Decentralization and diversification were the foundation for the Commission's business pluralism. A system of more numerous broadcast stations restricted to local service areas was preferred to regional stations, fewer in number but with more powerful transmitters. A decentralized communications system entrusted to local business ownership, the FCC reasoned, would have the built-in structural advantage of being more accountable and thus more responsive to local public interests and needs. Commission rhetoric suggested a strong opposition to monopoly. The goal of diverse and antagonistic media voices demanded not only a multiplicity of local owners, but independent local media as well. Diversification required that local broadcasters not be owners of other mass media, like newspapers, which serve the same locale.

Because the FCC eschews performance appraisals of broadcasters' public responsibilities in license renewals, the renewal process has become an automatic and empty ceremony. In effect the Commission has chosen to deal with the question of the public's informational interests and needs in its initial licensing or entry policies. The agency has taken the position that in controlling station ownership from the outset, it can enhance licensee responsiveness to public interests. In this fashion democratic communications objectives can be achieved indirectly through ownership controls, thus avoiding the chilling effects on free expression which would accompany intervention in the area of programming policy. Under such logic has the public's stake in the electronic information structure been entrusted to the efficacy of a presumed business pluralism.

In this chapter we will examine the fate and efficacy of local ownership in American television. The following chapter will consider the diversification policy of media independence.

THE HISTORY OF LOCAL OWNERSHIP

In the past decade, locally-owned television stations have become commercial curios rather than the industrial norm. But the

FCC's faith in local ownership, as the foundation of business pluralism, has survived aloof from and unruffled by the development of network and group broadcasting. A comprehensive 1965 policy statement offers an example of FCC reaffirmations in this regard.

The document begins with a ritualistic reminder of the Commission's primary licensing objectives: "first, the best practicable service to the public, and second, a maximum diffusion of control of the media of mass communications." In Commission thinking, these two are cognate. The policy proclamation singles out the factors of diversification of control and local ownership as being the chief instruments to secure these objectives. The statement reads, in part

> Full-time participation in station operation by owners:
> We consider this factor to be of substantial importance.
> It is inherently desireable that legal responsibility and
> day-to-day performance be closely associated. In addition, there is a likelihood of greater sensitivity to an
> area's changing needs, and of programming designed to
> serve these needs, to the extent that the station's proprietors actively participate in the day-to-day operation of
> the station.
>
> The discussion has assumed full-time, or almost full-time, participation in station operation by those with
> ownership interests. We recognize that station ownership by those who are local residents . . . may still be
> of some value even where there is not . . . substantial
> participation. . . .[7]

Here local ownership is presented as a means to grassroots business management. That is, the Commission prefers managerially active local owners; but it will favor local ownership per se, in any case. The policy thus increases the probability of an integration of ownership and control. Stations with such a grassroots base in the community are expected to be more inclined to serve the communications needs of local publics.

This follows from several assumptions. First, owners who are local residents have a stake in their community and its welfare. Accordingly, they should thereby be motivated to provide programming which will aid the community in resolving matters of social, political and economic concern. Second, local ownership implies familiarity with local social and economic conditions, with the particular needs of business, labor, educational, racial, ethnic, and other community groups, and familiarity with representative leadership. Third, it is assumed that an established member of the community is in a better position to provide a full range and balance of views on local concerns and issues. Finally, a local owner with strong ties to the business, professional, civic, and social affairs of the community is likely thereby to be held accountable to local needs and interests, and to fulfill programming promises and commitments.

Obviously, these assumptions do not apply to absentee owners, the corporations which have, through construction or purchase, assembled a chain of stations under their control. No one imagines that the institutional investors, founding families, and widely dispersed small shareholders which own groups of stations, would qualify for grassroots status. These multiple, or group, owners constitute the opposite of the Commission's locally owned and operated ideal. Any group ownership, by definition, erodes business pluralism. First, because it reduces the number of owners licensed to broadcast from the optimum that would exist if each television station were separately owned. Second, because it refines the organizational means for monopoly power over the information channels by a few business interests.

Thus the objectives of the local ownership policy are twofold: first to maximize the probability of business responsiveness to local public communication needs and interests; second, to maximize antimonopolistic constraints. Through such a policy, a pluralistic, decentralized community influenced communications system could be realized within the framework of private business ownership. Public interest responsibilities would be built into the system.

The Commission, it would seem, should be commended for its ingenuity in discovering a simple principle which both satisfied public interest requirements and abrogated the need for government intervention in broadcast operations.

But philosophy and practice are not often equivalent. A clue to their contradiction here may be found in the regulatory traditions of the FCC itself.

The legislation which created the FCC in 1934 was in its essential respects a faithful reproduction of the first comprehensive regulatory statute to concern broadcasting, enacted seven years before. The Radio Act of 1927, communications historian Sydney Head tells us, "was to a large extent the product of the radio industry itself."[8] That Act established the Federal Radio Commission. During the 1920s, the development of radio had been shaped by internecine struggles within the broadcast patent cartel that the Radio Corporation of America, General Electric Corp., Westinghouse Electric Corp., and the American Telephone and Telegraph Company fought to maintain; pirate broadcasting by amateurs and others who ignored the cartel's patent privileges; and a hands-off, laissez-faire posture by the federal government. In 1926 the chaos that these automatic free market mechanisms had produced reached crisis proportions. The overlapping use of frequencies often made radio listening a disturbing experience—so much so that the manufacturers suffered a market decline in receiver set sales. Consequently the industry turned to the government for its assistance in imposing some order on the marketplace.

The origin of the regulatory function was not fundamentally to make broadcasting's economic forces submit to the control and communications interests of the public. Rather it was to guarantee sufficient stability in development to enable the developers to prosper without fear of cutthroat competition. The Federal Communications Commission, like its predecessor agency, practiced promotional regulation. Faced with demands that the society's legal institutions foster the investment of capital, labor and management in developing this new technology, the Commission provided the legally sanctioned order necessary for market operations. Frequency allocation, station licensing, and other regulatory procedures created the necessary frameworks of expectation within which private investment and profit-taking could thrive. Promotional regulation enabled first radio and later television owners to know in what respects law permitted them to impose their will on others (programming "freedom"), and in what respects law protected them from the actions of others (controlled competition and shelter from government interference in operations). In superintending the distribution of power in broadcasting, the FCC provided for the public's interest chiefly by legitimating and supporting the designs of private business.

More often than not, this harmony of interests between the FCC and the communications industry drowned the lyrics of the agency's professed goals, and the local ownership policy was no exception.

The history of the local ownership concept predates the development of television broadcasting. It was first proposed as early as 1928 by the Federal Radio Commission, and was subsequently incorporated by the FCC to apply to ownership of radio facilities.[9] Following the practice that developed in the course of radio regulation, the Commission early initiated a series of compromises with the local ownership doctrine at the advent of television. In 1941, the same year that the Commission first began licensing commercial television stations, it also began eroding its local ownership policy. It did this by adopting the first TV multiple ownership rules. These permitted common ownership of three television stations. The rules still specified the 'one license per owner' ideal, stating

No person . . . shall, directly or indirectly, own, operate or control more than one television broadcast station, except upon a showing (1) that such ownership . . . would

foster competition among television broadcast stations
or provide a . . . service distinct and separate from existing services, and (2) that such ownership, operation
or control would not result in the concentration . . . in a
manner inconsistent with public interest. . . .[10]

But the litany of exceptions revealed the Commission's penchant for
stimulating the growth of network television. Business pluralism
would be accommodated to absentee group ownership, where the opportunity for distinct services, meaning networking, existed. With TV
broadcasting in its infancy, the Commission responded pliantly to the
claims of industry leaders that "chain broadcasting" (as networks
were then called) required chain ownership. The eagerness of the
Commission to patronize the designs of network entrepreneurs proved
increasingly bold: in 1944, with only six commercial stations on the
air the FCC raised its three station limit on common ownership to
five.
     The marked increase in the number of stations and license ap-
plications following the Second World War convinced the Commission
that its initial allocations limiting television broadcasting to the VHF
spectrum would place an unnecessary ceiling on the number of available channels. It therefore ceased granting initial licenses in 1948 to
provide time for preparing a revised plan of spectrum use. The freeze
was lifted in 1952, and the Commission announced plans to expand the
commercial spectrum to include UHF telecasting and eventually inte-
grate VHF and UHF bands. The VHF spectrum could have accommodated only a maximum of 650 stations for the entire nation. When it
resumed licensing in 1952, the FCC unveiled its first comprehensive
allocation of channel assignments providing for 2,053 stations, both
VHF and UHF, spread over nearly 1,300 communities.
     Embarked upon in a vacuum of supporting laws and provisions,
the policy was destined to fail from the start. The four year freeze
had entrenched preexisting VHF stations with both viewers and advertisers. The intermixture provisions had set poorer signal quality UHF
telecasters in competition with the farther traveling signals of VHF
stations in the same community. In the absence of laws requiring all-
channel set manufacture, few households had sets which could receive
UHF. With UHF stations everywhere going dark almost as quickly as
they were authorized to broadcast, the FCC, faced with the failure of
its expansion policy, turned reflexively to its ownership policies for a
means to lead prospective UHF investors out of the dark.
     The 1954 Report and Order announcing the Commission's re-
vised strategy is a sobering example of bureaucratic ratiocination.
Shrouded in a lengthy discourse reaffirming the salience of diversification and local ownership policies, and disclaiming any departure

from previous Commission philosophy, the Report goes on to raise once again the limit on common ownership from five to seven stations.[11] As one who studied the early years of FCC operations noted,

> it is not uncommon to give the rhetoric to one side and the decision to the other. Nowhere does the FCC wax so emphatic in emphasizing public service responsibility, for example, as in decisions permitting greater concentration of control . . .[12]

The multiple ownership rules were revised ostensibly to permit group owners to acquire two additional UHF stations. By yielding on multiple ownership, the Commission hoped to stem the rapidly curtailing development of UHF telecasting. Group owners, reasoned the FCC, would be better able financially to nourish new UHF stations and endure their presumably temporary economic hardships. The new rule still limited to five the number of VHF stations commonly owned.

Again local ownership was the casualty, expansion and development the cause. The bitter irony was that UHF development had been envisioned as a means to expanded competition, diversity, and an enhancement of business pluralism generally. Instead monopolies were encouraged to multiply, but in the name of competition.

Two years before this last rule revision, Congress had approved an amendment to the Communications Act which the industry had lobbied for stubbornly.[13] The amendment gave the FCC new orders to follow in passing on station sales.

Originally the Commission had followed a policy of accepting or vetoing a sale, with the buyer of a station being chosen by the incumbent owner. Then, after 1945, the FCC maintained that it could itself make the choice among prospective buyers of a station. This meant that the Commission might not always select the highest bidder.

Alarmed by what its trade press called "an assault upon the time-honored concepts of the rights of property owners," the industry first persuaded the Commission to abandon this competitive selection policy in 1949.[14] Then broadcasters shifted their focus to Congress, where they asked for an amendment which would make the competitive sales process illegal.

The amendment institutionalized the auctioning of stations to the highest bidder, which typically turned out to be an existing or would-be group owner. While shopping around for local bidders had never been a Commission practice, the new law worked to make difficult any future FCC interference in sales of locally owned stations to group broadcasters.

As a result, the multiple ownership rules themselves began to acquire new significance as a means of actually limiting group ownership from still further expansion. Ownership trends encouraged this. By 1956, with over 450 TV stations on the air, locally owned stations accounted for just 55 percent of the total. Groups already controlled almost half of the licenses. Gradually, then, the conflict over owner- ship rules began to shift away from efforts to increase the station limit in the face of growing pressures to reduce it.

The pressure to roll back the rules began to mount in 1956 when the federal district court charged that there was "some suggestion that the Commission has changed, or is changing, its view as to the dominant importance of local ownership and as to the evil of a concen- tration of the media of mass information."[15] A similar conclusion was reached by a special Network Study Staff, authorized by Congress, which released its report in 1957.

The report began by scolding the FCC, noting that "the com- munity institution concept has been seriously eroded." The Study Staff stated it had found "little evidence" that the earlier rule re- vision had resulted in significant development of UHF broadcasting by group owners. To correct this it urged "further limitation, rather than relaxation," of the multiple ownership rules, and called for (1) a long run goal of one station per licensee; (2) an initial licensing pre- sumption in favor of local resident ownership; and (3) immediate adop- tion of a three VHF station limit on group ownership in the top 25 markets, allowing multiple owners three years to sell stations in order to conform to the rule.[16] However, the Commission chose to ignore its staff report, for a time.

In 1961, a new chairman, Newton Minow, was appointed to the FCC by President John Kennedy. In his inaugural address to broad- casters assembled for their annual convention in Washington, Minow characterized television programming as a "vast wasteland." He praised several public affairs offerings of the networks, but found them excessively rare. Minow told broadcasters pointedly "Never have so few owed so much to so many."[17]

While Minow's tenure at the FCC proved to be brief, after three years he left behind a sense of activism, which was novel for the Com- mission.

Meanwhile, the number of television stations locally owned had continued to shrink, from 55 percent in 1956, to 43 percent in 1964. Group owners now controlled a majority of licenses.

That year the Commission responded with a new interim policy designed to discourage further multiple ownership in the fifty largest markets. The policy required that any owner of one or more VHF stations in the top fifty markets who made application to buy another VHF in those markets would have the application automatically desig- mated for hearing.[18] This seemed to be a guarantee of added red tape, something the Commission hoped would frustrate group purchases. But within a month, a swift protest from group owners, in the form of a petition for relief, induced the FCC to retreat from its position.

The following year the FCC again proposed rule amendments to protect more locally owned stations from being sold to group owners. The Commission chose a recommendation from its staff report, which by then had been awaiting action for eight years. It slated for hearing a rule prohibiting group owners from acquiring more than two VHF's, and a total of three television stations (VHF and UHF) in the top 50 markets. Simultaneously a revised interim policy was adopted which required hearings on any ownership transfers that would result in group ownership exceeding this proposed standard.[19]

In three years of deliberating the proposed rule, the interim protective policy was overruled 16 times, as the Commission routinely approved group purchases which exceeded the standard under consideration. This evidently defunct interim policy and the proposed rule were both dropped by the Commission in 1968, amidst voluminous filings from group owners urging rejection.

In the course of these deliberations the number of TV stations locally owned had decreased again, from 36 percent to only 26 percent.

Table 1 shows the fate of local ownership from 1956 to 1974. Groups now own a heavy majority of stations in all market categories. But group ownership rises (and thus local ownership diminishes) increasingly as one moves to the largest, most profitable markets. The top 25 markets, accounting for 50 percent of television's weekly audiences, show the highest concentration of group owners.

Much but not all of this decline in local ownership can be blamed on the Commission's willingness to approve station sales by local owners to group interests. The Commission seems no less reluctant, however, to permit existing group owners numerical increases by constructing new stations. Changes in the ownership structure of the top 50 markets are instructive. In the four years from 1966 to 1970, group owners purchased eight stations from local interests, while no group stations were turned over to local owners. During this same period FCC authorizations enabled local applicants to start 15 new stations in these markets, while 18 other new stations were started by groups.

Clearly, a policy that is so casually and prematurely relaxed, as this one was in the 1940s when rule changes were solicited by network aspirants; that is so easily made the scapegoat, as this was in the 1950s when lagging UHF development prompted an inducement to monopolization in the name of competition; and which provokes such protracted deliberation when enforcement is threatened, as was the case in the 1960s when efforts to salvage the policy by strengthening the rules dragged on unsuccessfully for four years—clearly such a policy is more a philosophical artifact than a structural control device safeguarding the public interest. But if it is fair to say that the

TABLE 1

Local Ownership of Television Stations

	1956	1966	1974
Top 25 Markets			
Stations Broadcasting	90	109	136
Percent Locally Owned	28.8	23.9	20.5
Top 50 Markets			
Stations Broadcasting	162	193	230
Percent Locally Owned	43.9	30.6	27.0
All Markets			
Stations Broadcasting	457	592	693
Percent Locally Owned	55.2	33.2	32.2

Source: Television factbooks, broadcasting yearbooks.

local ownership policy is often proclaimed, but almost nowhere enforced; that over the years the Commission has been given more to monitoring changes in the business structure, than to mandating them; and that business pluralism, if it exists at all, must be an ancillary prophylactic, not a core characteristic of television's ownership structure; if these are fair appraisals, it does not, however, follow that the precarious and infrequent occurrences of local ownership in the business structure invalidate the premise of business pluralism. For while the phenomenon may be obsolete, the premise may be valid. In that case, whatever inadequacies may exist with regard to a diversified and democratic communications flow might be more reasonably attributed not to the liberal theory of communications control at issue here, but to its practice.

However, before examining the premise of business pluralism any further it is useful at this point to consider some general characteristics of the communications flow as it has been structured under business ownership, dominated by station groups.

SURVEYING THE EFFECTS OF BUSINESS OWNERSHIP

Beginning in 1968, a series of studies appeared which indicated that local informational programming on television was often meager. They showed that public affairs programming, the informational format which was more likely than others to afford opportunities for the expression of diverse and antagonistic views, was often nonexistent. The first of these was an independent analysis of television stations in Oklahoma by two FCC Commissioners, Kenneth Cox and Nicholas Johnson. Their survey of television programming undertaken independently of the Commission indicated that "with a few exceptions, Oklahoma stations provide almost literally no programming that can meaningfully be described as local expression." They concluded that, in Oklahoma at least, "the concept of local service is largely a myth."[20]

The Commissioners repeated their study the following year, surveying all television stations in the state of New York.[21] This time they employed an expanded and more detailed quantitative analysis of programming. Their findings suggested a wide range of programming performances, varying from station to station. In the estimation of Cox and Johnson, some stations showed a considerable amount of local informational programming in both news and public affairs formats, relative to other stations. Stations which allocated higher percentages of broadcast time to local information tended to be in New York City, the nation's largest television market. Higher percentages were found particularly among the network-owned and -operated flagship stations, where the network national news staffs are headquartered. A majority of other stations evidenced exceedingly poor performances in this respect.

A third study which incorporated these methods was issued in late 1969, surveying stations in the Middle Atlantic region (the Washington, D.C.-Maryland-Virginia-West Virginia area).[22] This research was conducted by audience groups, and it concluded with the recommendation that the Commission investigate seven of the 32 stations in the survey because of serious deficiencies in informational programming and other selected broadcaster responsibilities. For example, they found that of the total air time, the amount of programming dedicated to public affairs on these stations ranged "from very low (a maximum of 6.3 percent) to virtually nonexistent (.27 percent). . . ."[23] By comparison, they found that the amount of air time devoted to commercials in a sample of these stations ranged from 9 percent to 18 percent.[24]

Together these case studies, which include several of the nation's largest television markets, provide some evidence that a substantial number of stations have demonstrated little willingness to provide public forums, and to communicate social controversies and conflicts. At least the absence of programming that would constitute local expression, which Cox and Johnson found in most cases, points to this. And one can infer little willingness to service diverse and antagonistic interests in the communications flow from the Middle Atlantic study finding that business interests and commercial needs were allotted far more broadcast time, for a fee of course, than were all the other public affairs of the audience.

As was pointed out in the previous chapter, in our centrally managed information system, absent access rights and the nature of the communications inequalities that result, are defined on a day-to-day basis by that system's willingness to function as a public forum. The willingness to service free expression by those interest constellations outside of the business sector, by those groups disenfranchised with respect to communications control, is the main ameliorative mechanism in what is otherwise an authoritarian information flow.

Of course measures used above, such as the proportional amounts of public affairs programming, fail to discriminate between broadcasts which provide a full panoply of diverse views on emergent and penetrating social conflicts, and those narrowly conceived programs which seem to pass superficially over superficial topics of doubtful substance—not to mention the qualitative shadings lying between these extremes. A recent study of broadcasters' conceptions of controversy presents some sobering findings in this regard.[25] In this survey (financed by the National Association of Broadcasters), stations which responded that they had aired programs addressing "politically controversial ideas" were asked to list specific examples. The 1966 list of 42 topics included unidentified flying objects, daylight-saving time, uterine cancer, and others of a similar genre. The 1971 list of 28 topics included rock music, long hair, and the high cost of funerals. The researcher's only generalization from his findings was that "it was doubtful that controversy properly understood was all that prevalent in news and public affairs programming, including editorials, in 1966 or 1971."[26]

The public-forum functions of news broadcasts are even more limited. Quickly paced capsule items, which in the print medium would barely fill a front page, are interlaced with commercial messages and promotional announcements and are packaged like the commodities they are designed to sell. The news shows which result often seem to narrow the distinction between information and entertainment. The "happy news" format, in vogue on some stations, in simply a more conscious exaggeration of this. All of which suggests that news formats, generally, would seem to be less efficacious than public affairs productions in providing opportunities for diverse free expression.

These qualifications about the forum functions, and many others that could be made about news and public affairs programs offered by the present commercial system, simply acknowledge an obvious truth about communications systems that enfranchise some with privileges and others not at all; namely that control over communications carries with it the power to determine the manner in which information is presented and the way in which controversies are defined and discussed.

In an attempt to extend the investigations initiated in the case studies referred to above, a national survey of station programming was undertaken by this writer. The findings reported below hold several advantages over similar surveys presently available.

First, the sample is national and not regional, and the strict selection process required in random sampling was applied in order to accurately reflect any variations in station programming practices. The survey is focused on television's 75 largest markets, which include 75 percent of the national weekly viewing audience. One third of the total number of stations in these markets was sampled.

A second advantage is that the programming data itself comes directly from local editions of TV Guide. This permits independent categorization of programs. The earlier studies of Cox, Johnson and the audience groups used program totals categorized and calculated by individual station managements and furnished to the FCC. Figures prepared by interested parties are considered, quite appropriately, to be more suspect and less reliable than independently classified tabulations. For that reason the stations' own reports were not used in this survey, and TV Guide was found to be the only publicly available alternative.

Fewer than 1 percent of the programs listed and described in TV Guide presented any difficulties with respect to their appropriate classifications. These few programs were classified only after oral or written content descriptions were obtained from the respective stations.

Table 2 summarizes the findings of this investigation. The table shows the mean percent of total broadcast time stations allocated, first, to all public affairs programs; and second, to those public affairs programs which were produced locally by the station.* The table

---

*The definition of what constituted a public affairs program was the same as that authorized by the FCC, except that editorials were not included. In its license renewal application (Form 303), the Commission says "public affairs programs include talks, commentaries, discussions, speeches, political programs, documentaries, forums, panels, round tables, and similar programs primarily concerning local, national and international public affairs."

TABLE 2

Television Informational Programming
in the Top 75 Markets, 1970

Mean Percent of Broadcast Week	Programming			
	Local public affairs	All public affairs	Local news	All news
VHF, N = 98	0.9	2.1	6.9	10.3
UHF, N = 27	0.6	1.6	3.3	5.3
Network Affiliates, N = 103	0.9	2.1	6.7	10.4
Independents, N = 22	0.8	1.4	3.6	3.6
All Stations, N = 125	0.8	2.0	6.1	9.2

Source: Compiled by the author.

also reports the mean percent of air time devoted to all news programs combined, and to locally produced news.

The results probably confirm the impressions built up by most
viewers of the commercial medium. While news programs were allotted slightly more than 9 percent of air time on the average station,
public affairs offerings were virtually nonexistent. The news programs
in these calculations include the commercials within the program; and
for local news, weather and sports segments are included as well.

Most likely to approximate the forum function are the public affairs productions, and these, it appears, are also the programs most
likely to be avoided by broadcasters. The communication of local
issues, controversies and viewpoints through such formats was unusually uncommon, occurring less than 1 percent of the time. The 0.8
percent figure, to be exact, is equivalent to only one hour of public
affairs offerings in a week which, for this sample, averaged 124 hours
of broadcasting. More than a quarter of the 125 stations surveyed
produced no programming whatsoever in this category. The most produced by any station was 5.2 percent.

Adding in nationally produced network and syndicated public affairs broadcasts hardly improved the picture, raising the total slightly to 2 percent, or two hours and 30 minutes per week.

In both the locally produced news and total news programming categories, network affiliation and spectrum frequency made a difference in station performances. VHF stations, on the average, almost doubled the news programming of UHF stations (10.3 percent compared to 5.3 percent); the same differences held in the case of network affiliates versus independents. These findings are consistent with economic pragmata that are well understood in the industry, and that are reflected in annual financial reports of the FCC: that VHF and network-affiliated stations are both consistently more profitable and amass greater revenues than either UHF or independent stations.

One can explain these consistencies in several ways, however. A recent study reiterates an argument that is frequently summoned by broadcasters and their representatives to the effect that more affluent stations can better afford costly news programming.[27] The implication is that news is unprofitable, and that affluence breeds the altruistic, public-spirited broadcaster. For the present it is sufficient to note that on the contrary, news is indeed profitable, as broadcasters themselves attest; and that it is very popular, not only among audiences, but among advertisers.[28]

Rather than invoke altruism, one might more persuasively explain the coincidence of affluence and greater news programming by means of some additional economic considerations. As was pointed out several years ago in a trade-press article, "The maintenance of a news organization is costly, but after the initial outlay, any expanded uses of that organization brings higher returns at a minimum of increased costs."[29] Since the more affluent VHF and network affiliated stations can more readily afford the initial costly outlays necessary to move beyond "wire service" news programs, one could expect that they would spread the costs over longer broadcast hours and take advantage of the resulting accelerating profits.

If altruism were in fact operating, one would expect it to be manifested in public affairs programming especially, since documentary and forum formats are more clearly related to public service responsibilities. But here performance differences between affiliates and independents, VHF and UHF stations are negligible.

A pervasive component of the electronic information structure not shown in this table is the commercial message. At the time of this survey, the National Association of Broadcasters' Television Code permitted stations to allot commercial interests about 27 percent of their air time. The only exception to this was prime time, (which the Code defined as any three consecutive hours between 6 p.m. and midnight). During these few hours, commercial messages were limited to 17 percent of air time.[30] Of course not all broadcasters were able to sell message time to commercial interests for 27 percent of every hour they were broadcasting. This fact is reflected in the data on commercialization reported in the Middle Atlantic study which showed that for a full week, commercial time on selected stations ranged from 9 percent to 18 percent of broadcast time. But occasionally broadcasters exceed, during some hours, the 27 percent of air time that their own trade association has established as the commercial maximum. Among stations included in the survey profiled in Table 2, for example, 38 percent reported to the FCC that in the course of a typical week they exceeded the 27 percent ceiling on one or more occasions.*

Comparing the commercial flow to the public affairs flow reveals some interesting characteristics about the electronic information structure generally. Several factors are involved in this comparison.

First, commercialization represents that portion of broadcasting which is given over to financing the production costs of all other messages in the system. Thus in looking at the system as a whole, it is important to ascertain what the public expression and forum dividends are relative to the costs. Of course commercialization does not reflect the full costs of the system of mass communications, which would also have to include the technological hardware—both transmitting facilities and receiving devices. But the public pays directly 96 percent of these charges, since the ratio of receiver set investments to investments in broadcasting plants and facilities is more than 20 to 1.[31] Thus commercialization represents a crude but useful measure of the system's message costs, with public affairs programming a reflection of dividends.

Second, commercialization provides a form of direct, albeit discretionary, access to the public for society's business interests. Competing or conflicting nonbusiness interests can secure discretionary access chiefly through public affairs formats, since broadcast advertising time is not regularly available to any except business interests. Both the 60-second commercial spot and the 25 minute press interview or documentary have their distinctive limitations, but they share the common function of limited access.

---

*This information was obtained from license renewal forms filed triannually by the stations in the sample, and includes programming periods during 1969, 1970 and 1971. The Commission designates a typical week and requires programming information from the stations for that period.

Third and finally, commercialization is one measure of the degree to which the present communications flow services business affairs, analogous to the way panel discussions and documentaries
service public affairs.

Unfortunately, comparable data for public affairs programming
and commercialization are not readily available for the national station survey reported on here. Independent tabulations of station commercialization practices would require overwhelming monitoring
efforts, or comparably tedious surveys of station program logs. But
in license-renewal forms filed with the FCC, most stations included
in this sample have indicated their commercialization practices for a
typical week, consolidated into gross categories. This data shows for
each station, in the course of a week, the number of hours broadcast
with commercials occupying more than 20 percent of air time; the
number of hours with commercials occupying from 13 percent to 20
percent of air time; and the number of hours with commercials occupying less than 13 percent of air time. For the sample as a whole,
stations averaged more than 20 percent commercialization during 24
out of every 100 hours broadcast. They averaged between 13 percent
and 20 percent commercialization for 31 out of every 100 hours broadcast. And during the remaining 45 hours of every 100, they averaged
less than 13 percent commercialization.

While these figures cannot be neatly reduced to a grand mean,
comparing these commercial percentages to the average public affairs
percentage of 2 percent suggests several general characteristics about
the communications flow as it has been structured under business
ownership.

First, the public expression and forum dividends in telecommunications are quantitatively insignificant compared to the proportion
of broadcast time assigned for financing the operation of the medium.
To be simplistic, the costs are exorbitant when compared with the
dividends. Second, business interests enjoy far more discretionary
access to the television system than do other interest groups. Third,
the system functions to service business affairs excessively when
compared with the marginal to nonexistent servicing of general public affairs.

In effect these characteristics simply restate from several per-
spectives the conclusion evidenced by the national data reported in
Table 2: that the owners of the existing telecommunications system
show a general and pronounced unwillingness to have the electronic
medium function as a public forum, and to service free expression by
those interest constellations outside of the business institutions.

Yet if the record here shows a general unwillingness in this regard, might it not be because the ownership structure is characterized
more by competing monopolies than by the business pluralism the FCC

envisioned as operative? In other words, are the present deficiencies in the communications flow primarily attributable to failures in implementing the philosophy of local ownership? Are they attributable to the fact that the present structure is managed primarily by group owners? Further, can such deficiencies be corrected within the framework of a business-controlled communications system, one characterized by business pluralism? This brings us full circle to our initial problems: is the promise of local ownership, and hence business pluralism, valid?

LOCAL OWNERSHIP: BUSINESS AS USUAL

Not surprisingly, it is the group owners themselves who are first to dispute the grassroots accountability assumptions about local ownership. And they do so not by retreating to abstract formulations, as the FCC is prone to do in describing its ownership policies as "inherently desirable," their underlying logic being "self-evident." Instead, little is spared financially by the group owners in procuring respectable academicians and private consultants to make the argument for them. For a handsome fee, these technicians lend their legitimacy to self-serving methodologies and analyses which result in thick volumes of argument, selectively buttressed with research and data consistent with the industry stand.

The FCC's hearings on strengthening ownership rules in the mid-1960s offer a case in point. In 1966, a hastily formed consortium of group owners produced a study from several Harvard professors dealing with the effects of group ownership, for which they paid an estimated $300,000. With regard to the question of pluralism and diversity, the report concluded that "the program activities of group-owned stations are closer to the FCC's goals of program diversity than are the similar activities of single owners."[32] However, a careful reading of this 200 page report produces little evidence to support such conclusions. The chapter on which the claim quoted above was based shows group and locally owned stations to differ in only two respects: (1) group stations on the average broadcast 40 minutes longer each day than nongroup stations; (2) group stations allocated on the average 11.3 percent of their broadcasts to local programs (entertainment and informational programs combined), whereas nongroup stations allocated 10.0 percent, a difference of 1.3 percent. The authors extrapolated these findings to mean that group ownership contributed more to diversity in communications than local ownership. Within a few years this report was being cited by other industry-financed studies as having "demonstrated that multiple ownership facilitated

independent and public service programming, and was in several respects favorable—not detrimental—to the national public interest."[33]

Even in periods unmarked by proposed ownership rule changes,
industry-financed studies surface which, no matter what their particular findings, conclude by reaffirming the virtues of a group-dominated
ownership structure. A recent example is a research monograph by
Professor Frank Wolf, which was financially supported by the National
Association of Broadcasters.[34]

Wolf's study reported a new survey of news and public affairs
broadcasts, geared to relating ownership differences to quantities of
informational programming. His findings showed that television public affairs programming had decreased between 1966 and 1971. And
in both years group owned stations aired fewer minutes of informational programming (news and public affairs combined) than single
owner stations. Specifically, groups averaged 16 fewer minutes per
day in 1966, but only seven fewer minutes per day in 1971. However,
in response to Wolf's questionnaire, group owners claimed to have
allocated a slightly higher percentage of broadcast time to informational programming than single owners claimed to have allocated.
Wolf also reported that the six network-owned stations he surveyed
averaged more minutes of informational programming per day than
single owner stations: 20 more in 1966, but only ten more in 1971.[35]

Since Wolf failed to calculate percentages for his own independently measured data, the discrepancies between his findings (based on
absolute figures) and the owners' claims (expressed in percentage
estimates) were not resolved in the study. The findings were further
complicated by Wolf's use of a very small sample, comparing just 14
group-owned stations to only seven single owner stations; and by his
decision to choose a nonrandom sample, which resulted in acknowledged sample bias.

While the data constrained Wolf to state that group-versus
single owner differences in quantities of informational programming
were "negligible," he nonetheless concluded, without further supporting evidence, that there were "greater news-gathering resources
(that is, economic resources) available to station groups, as compared
to stations owned by parties with no other television interests."[36]
And whereas a difference of seven minutes in informational programming between group and single owners was "negligible," a difference
of ten minutes between single and network owned stations seemed to
him to be "substantial."[37] He somehow concluded that in these findings just reported, "it was shown that the richer and more profitable
the station or network, the more likely it was that increased amounts
of news and public-affairs programming would be shown. . . ."[38] As
if the message was not yet clear, Wolf went on to note

Meanwhile, the FCC was proclaiming its determination to encourage conditions of increased competition in the industry and increased diversification of ownership of television stations. By implication, this objective would require the number of single-owner stations to mount, the number of unaffiliated stations to increase, and, in general, a decrease in the domination of the industry by the three networks.

Such developments, <u>in the absence of direct program regulation by the FCC,</u> would assuredly reduce the amount of news and public affairs programming seen, reduce the number of stations editorializing, and, probably, increase the dependence of individual stations upon advertisers, with a consequent increased aversion to controversy on the part of station management.

. . .

Another possible implication of the study is that those at the FCC and outside who hoped to increase the quantity of news and public affairs programming on commercial television should abandon their efforts.[39]

It should be enough merely to note the abyss between findings and conclusions in the Wolf and Harvard studies above, to perceive the strain toward conformity with the approved industry line. But these studies, and others like them, are worthy of particular attention because of the very real function they provide in the world of government-industry relations. Essentially they are the ideological briefs of the industry pawned off on the regulators as technocratic intelligence, in order to give bureaucrats some justification for protecting industry interests. They make the job of Commission rationalization easier, in the pejorative sense. Otherwise the FCC could investigate these matters itself. But the Commission almost never does studies of its own, preferring instead to invite comments from interested parties.

In themselves, both the Wolf and Harvard studies attest to an impoverishment of evidence which could win approval for the prevailing ownership structure. But they are ironically conceived to demonstrate its virtues.

The larger question remains: Are the local owners, banished to the recesses of a network- and group-dominated system, demonstrably more willing to service public affairs? A preliminary study by Baldridge of 29 network-affiliated stations in ten selected markets indicates that single owners are no different than group owners in number of weekly hours spent on local public affairs, syndicated public affairs, and news, or on any of the other program categories he investigated, for that matter.[40]

This writer's data, reported below, corroborates the Baldridge findings. And with a random sample including affiliated and unaffiliated, VHF and UHF stations over the largest 75 markets, confidence in the data as an accurate representation of industry structure and performance should be enhanced.

Table 3 gives no suggestion that the FCC's ideal form of business ownership, the local owner, is particularly willing to provide a forum for local affairs, or for public affairs generally. The local owner averages only six-tenths of 1 percent of total air time in the local public affairs category, and less than 2 percent in overall public affairs. If such programming is the principal means of servicing free expression by those not owning television stations, then the amount of such expression is indeed trivial, relative to other components of the electronic communications flow. The local owner offers no improvement in community responsiveness or in willingness to share his frequency privileges over the average station owner, discussed earlier.

The previous findings of deficiencies in the communications flow (see pp. 24-31) cannot then be blamed on the preponderance of group owners in the business structure. The local owner, like the group owner, is most likely to avoid that programming which is most likely to approximate a forum function in communications. In news programming, which makes only a limited contribution in this respect, the local owner is less constrained. But his allocations or air time do not differ from that of the group owner.

Comparatively speaking, the program differences by ownership category in Table 3 are all fractional. But in the case of local news, it happens that the differences are statistically significant,* suggesting that group owners actually do more local news programming, however marginal in terms of actual hours per week, than the local owners. This is only a manifestation, however, of more fundamental station differences within these ownership categories. Group owners generally purchase the more economically lucrative stations. Thus they tend disproportionately to be VHF- and/or network-affiliated stations. On the other hand, the local owners remaining in the industry tend disproportionately to be UHF and/or unaffiliated stations. These tendencies are reflected in the sample as well.

Controlling for this results in the percentages reported in Table 4. Only data for VHF network affiliates are presented since

_____

*At the conventional probability level of a 5 percent chance of error.

TABLE 3

Television Informational Programming of Group and
Locally Owned Stations in the Top 75 Markets, 1970

Mean Percent of Broadcast Week	Programming			
	Local public affairs	All public affairs	Local news	All news
Locally Owned Stations, N = 36	0.6	1.8	5.3	8.6
Group Owned Stations, N = 89	0.9	2.1	6.5	9.5
All Stations, N = 125	0.8	2.0	6.1	9.2

Source: Compiled by the author.

TABLE 4

Television Informational Programming of Group and
Locally Owned VHF Network Affiliates in the Top 75 Markets, 1970

Mean Percent of Broadcast Week	Programming			
	Local public affairs	All public affairs	Local news	All news
Locally Owned Stations, N = 23	0.8	2.2	6.7	10.5
Group Owned Stations, N = 64	0.9	2.2	7.3	11.1
All Stations, N = 87	0.9	2.2	7.2	11.0

Source: Compiled by the author.

there are too few stations in the remaining categories to permit meaningful comparisons in this sample. The fractional programming differences by ownership type seen in Table 3 disappear in Table 4 and the single statistical difference in local news programming is eliminated. Locally owned and group owned VHF affiliates show basically identical local public affairs allocations (0.8 percent and 0.9 percent) and overall public affairs time (each 2.2 percent). Local news hovers around 7 percent of broadcast time, with the 0.6 percent advantage group owners show in local news broadcast time accounting for an identical difference in total news.

CONCLUSION

We began this inquiry with the question, to what extent are the business corporations which control American telecommunications willing to afford opportunities for the free flow of ideas, opportunities which are critical to the practice of free speech and the freedom to communicate in an era of mass media. The legal structure, as we have seen, protects unequivocally the owners' right to free expression. In turn, it is asked to what extent these corporate managers provide for the rights to free expression of others—rights which the legal structure has transformed into privileges bestowed at their sufferance.

Note that the question is not whether telecommunications owners do in fact air diverse and antagonistic views, or only those views which conform to their interests. Rather there is the prior consideration, simply whether they evidence a willingness to provide the time for such expression.

The shape of the electronic information structure suggests they do not. It appears long on commercial information, and by comparison, relatively constricted on general news. But most important to free expression is the time available for public forums, for the conveyance of social intelligence of a critical nature. Here the flow of electronic communications is most repressed and constricted. In fact it hardly exists at all.

This probably comes as no shock—neither to those entranced by television, nor to others who intuitively find it a counterfeit mirror—yet the authoritative body in communications matters, the FCC, professes to have fused its two statutory objectives to aid broadcasting development and secure the public interest, without compromising free expression. While clearly playing an attentive and deferential role relative to business designs for developing and exploiting the spectrum resource, the Commission also gives assurances that its own particular ordering of the business structure will vouchsafe a responsive and accessible management and medium. That formula for ordering consists of a business pluralism, organized around the notion of independent, managerially active, local owners.

Increasingly as the structure has over time taken on a more centralized and monopolistic character, vocal minorities on the Commission and from within its staff have appealed for enforcement of this original ownership formula. Reacting to judicial criticism of regulatory failures to charges that television is a communications wasteland, and to restless audiences which mount license challenges, they straddle the apparent contradiction between exclusive control by business interests and a democratic interest in providing for free expression, whatever its source, by professing renewed faith in these pluralistic principles. And yet there is no evidence that even if enforced, they would make any difference in the sense of providing for a more accessible information architecture. In other words, business pluralism, in terms of the local ownership mechanism at least, gives no indication of ameliorating the tendencies of a centrally managed communications establishment to withhold from the larger society the potential for free expression. The communications establishment continues to reserve that privilege for itself.

NOTES

1. Communications Act of 1934 (U.S. Code 47), Section 153.
2. NAB President Justin Miller, quoted in Erik Barnouw, The Golden Web (New York: Oxford University Press, 1968), p. 232.
3. 74 FCC 1385 (1975). Television station WLBT in Jackson, Mississippi, in 1969 was denied renewal of its license for discriminatory programming against blacks, but not by the FCC, which had voted to renew the license. The renewal action was overturned in a landmark decision by the U.S. Circuit Court of Appeals for the District of Columbia. Office of Communication of the United Church of Christ v. F.C.C., 425 F 2d 543 (1969).
4. Jerome A. Barron, "Access to the Press—A New First Amendment Right," Harvard Law Review 80 (June 1967): 1645.
5. Associated Press v. U.S., 326 U.S. 1, 20 (1945).
6. Quoted in Barnouw, The Golden Web, p. 24.
7. Policy Statement on Comparative Broadcast Hearings, 1 FCC 2d, 394 (July 28, 1965).
8. Sydney W. Head, Broadcasting in America, 2nd. Ed. (Boston: Houghton Mifflin, 1972), p. 159.
9. FCC, Public Service Responsibilities of Broadcast Licensees (Washington: Government Printing Office, 1946), p. 12.

10. 6 Fed. Reg. 2282 (1941).

11. 11 P & F Radio Reg. 1519 (1954).

12. Murray Edelman, The Symbolic Uses of Politics (Urbana: University of Illinois Press, 1967), p. 39.

13. Communications Act of 1934, Section 310(b), amended July 16, 1952, 82nd Congress.

14. Broadcasting (June 4, 1945), quoted in Barnouw, The Golden Web, p. 227.

15. FCC Network Study Memorandum, "Multiple Ownership and Television," Journal of Broadcasting 1 (Summer 1957): 261.

16. Ibid., pp. 250, 261; and Network Broadcasting, Report of the Committee on Interstate and Foreign Commerce, House of Representatives, 85th Congress, 2nd Sess. (Washington: Government Printing Office, 1958).

17. Minow's speech is reprinted in: Newton N. Minow, Equal Time: The Private Broadcaster and the Public Interest, Lawrence Laurent, ed. (New York: Atheneum, 1964), pp. 48-64.

18. 64 FCC 1171 (1964).

19. Notice of Proposed Rule-Making, 30 Fed. Reg. 8166 (1965); Interim Policy Concerning Acquisition of Broadcast Stations, 65 FCC 548 (1965).

20. Broadcasting in America and the FCC's License Renewal Process: An Oklahoma Case Study, 14 FCC 2d 1, 12 (1968).

21. New York State License Renewals, 18 FCC 2d 268 (1969).

22. Ralph L. Stavins, ed., Television Today: The End of Communication and the Death of Community (Published for the Institute for Policy Studies; Washington, D.C.: Communication Service Corporation, 1969), pp. 89-187.

23. Ibid., p. 92.

24. This data on commercialization was collected for network affiliates in the Washington, D.C. area, and is thus only a sample from the stations included in the overall study; ibid., p. 185.

25. Frank Wolf, Television Programming for News and Public Affairs (New York: Praeger, 1972), pp. 94-98.

26. Ibid., p. 98.

27. Wolf, op. cit., pp. 126-36.

28. See, for example, Albert R. Kroeger, "News, News, News, News," Television 22, no. 2 (February 1965): 28ff.; Caroline Myer, "News Shows in Transition: Longer Forms, Different Hours, a Search for Individuality," Television 25, no. 7 (July 1968): 34ff.; and Television/Radio Age (September 21, 1970).

29. Myer, op. cit., p. 34.

30. Head, op. cit., p. 272.

31. See Herbert I. Schiller, "The Mass Media and the Public Interest," in Stavins, p. 59; and The Network Project, Control of

Information, Notebook Number Three (New York: March 1973), pp. 5-6.

32. United Research, Inc., The Implications of Limiting Multiple Ownership of Television Stations, 2 vols. (Prepared for the Council for Television Development; Cambridge, Mass.: 1966), p. I-1. The local-absentee labels are carefully avoided throughout this report.

33. George H. Litwin and William H. Wroth, The Effects of Common Ownership on Media Content and Influence (Prepared for the National Association of Broadcasters; Boston: 1969), p. I-2.

34. Wolf, op. cit.

35. Ibid., pp. 33-56.

36. Ibid., pp. 63, 139.

37. Ibid.

38. Ibid., p. 141.

39. Ibid., pp. 141-42.

40. Paul D. Baldridge, "Group and Non-Group-Owner Programming: A Comparative Analysis," Journal of Broadcasting 11 (Spring 1967): 125-30.

3

THE LEGACIES
OF CROSSOWNERSHIP

The local television medium arrived at an opportune moment in modern press history: the daily newspaper was fast becoming a local print monopoly, ostensibly because of sagging economic fortunes. Yet the number of American cities with daily papers was continually growing throughout this century, which seemed to dispute the claim that newspapers had encountered hard times financially. The financial changes that were occurring were of a different order: the number of cities with dailies operated by competing managements was shrinking drastically. In 1910, 57 percent of American cities with daily newspapers had competitive owners. By the mid-1960s, only 45 cities survived the mergers with a competitive press, representing just 3 percent of the total number of cities with daily papers. And half of the 45 communities with press competition remaining were left with newspapers managed under joint operating agreements, which strained considerably the definition of competition.[1]

When television appeared it was hoped that it might provide the opportunity for new, alternative sources of local expression. It was an instance of that syndrome of technological innovation referred to earlier. Just when the number of separately controlled channels of information in the daily print media was contracting under the press of monopoly ownership, technological wizardry produced a new distraction and a potential panacea. With television, the number of separately controlled information channels would, it was imagined, increase. Presumably this would mean a new pluralism in the management of the information structure, particularly so in the case of local communications, since television would be locally transmitted and owned.

From the outset regulatory policy for television was geared to capitalize on such opportunities. The policy of decentralization required not one or several national transmitters, but rather numerous local ones. Each local station's transmitting power was limited to prevent interference with other stations in nearby markets using the same frequency. Like other regulatory policies, the purpose here was to provide for independent outlets of local expression.

The FCC has frequently sought to increase the number of communications channels. That was its purpose in adding the UHF spectrum to commercial broadcasting in 1952, following the four year freeze in station licensing. In the belief that more UHF channels would be used if the multiple ownership rules were relaxed, those rules were amended two years later. When this failed to spur UHF development, the Commission in 1962 obtained a Congressional amendment to the Communications Act, empowering the Commission to require manufacturers to equip all new receivers to receive all 82 television channels.[2] While certain of these actions proved to be self-defeating, the Commission's stated intent was always to increase the number of independent sources of information and local expression. The envisioned decentralized pluralism, however, was always a business pluralism.

In the preceding chapter we considered business pluralism from the decentralization perspective, in terms of conflict between local ownership and monopolistic group ownership. Just as important to that pluralism, in the Commission's thinking, is the principle of diversification in ownership. The FCC has never used the diversification concept in a broad sense to refer to social diversities, such as those that might exist between profit- and cooperative oriented owners, or between business and labor ownership. Rather diversification has come to be used in a far more narrow sense to refer to a preference for a diversity of business owners between the different mass media: print and broadcasting. Thus it is tied in directly to the FCC ideal of plural, independent sources of local expression. For even if a television station is locally owned, if that owner is the local newspaper, then it does not truly qualify as an independent source of information and expression. Instead, the station simply adds another outlet to previously existing local communications sources. Accordingly, newspaper-television crossownership, as the phenomenon is called, is expected to reduce whatever pluralism exists between the media of mass communication, print and broadcast.

One can see how the FCC's focus on expanding the number of information channels, in the context of a decline in local newspaper competition, has made diversification an FCC synonym for opposition to crossownership.[3]

The Commission has, from time to time, also expressed disapproval of radio-television ownership combinations. But the newspaper-TV variant has, like the issue of group ownership, consumed much more of the Commission's energies in special investigations, licensing procedures, and most recently in hearings on rules which would ban local newspaper-television crossownership altogether. Because it is offensive to the FCC's ideal of business pluralism, and because the Commission has demonstrated a greater propensity to act in defense of business pluralism in this area than in the matter of group ownership, the significance of those local media monopolies bears some attention.

But if the Commission has been more aggressive, so too have the crossowners. The divisions that their pleadings have generated in regulatory circles complicate the issue. One Commission Chairman, Newton Minow, thought by broadcasters to be a utopian reformer, ironically has sought to advocate crossownership, while most crossowners themselves are content to merely defend it. The theory that most encourages FCC apprehensiveness about the wisdom of business pluralism is this—that the journalistic tradition of the print media has a salutary effect on broadcasting whenever ownership linkages between the two are permitted.

Thus contradictory claims surround the crossownership form of media monopoly. The traditional view is that only a strict pluralism, applied even to broadcasters who own local papers, will produce a more open and diverse communications flow. It would appear that this position is still persuasive to a Commission majority. The revisionist view, however, sees in newspaper crossownership precisely that kind of broadcasting management which will turn television from an entertainment medium to one which respects the informational and expressive needs of its audience. Whether either variant, the newspaper-free station, or the newspaper-owned station, evidences a greater willingness to provide forum opportunities and service informational needs than has been found to exist in the industry as a whole, will, therefore, merit attention.

HISTORY OF NEWSPAPER-BROADCASTING CROSSOWNERSHIP

Through the years newspaper-television crossownership has become solidly entrenched. It has not, however, like group ownership, exhibited steady growth.

The first application for a commercial television license was made by a newspaper, the Milwaukee Journal: in the early days of the medium substantial numbers of licenses the Commission awarded went to newspaper owners. In 1953, the first year after the four year freeze in licensing, newspapers owned 79 percent of the television stations on the air.[4] Newspapers quickly secured the choice outlets in the industry—those located in the largest cities and with the better frequency, VHF. Consequently, they easily won the most desirable network affiliations.

Some analysts read into these enterprising moves mainly altruistic motives: good will, prestige, and obligations to keep in step with current communications technologies are cited as reasons prompting newspaper license-seeking.[5] However, the reflexive invasion of newspapers into each new broadcast medium introduced in the twentieth century suggests a much different purpose.

With the introduction of AM radio in the 1920s, one media historian tells us that newspaper owners saw "a valuable publicity adjunct to boost newspaper circulation." Just as important, he continues, "radio was seen as a defense against possible changes in news reporting methods."[6] Accordingly, newspaper owners built radio stations, the better "to steer the new art along lines which would not interfere with the stability of their own field."[7]

Monopoly control provided one means to this end. One western publisher, owner of two dailies and two AM stations, was in 1930 "working on a plan . . . that will get us control of the air of the state."[8] Newspaper-AM ownership doubled and redoubled in the early 1930s, largely in order "to tailor radio into something the press could still handle," as AM had proved a considerable drain on newspaper advertising revenues.[9]

Just how the press could handle radio became clear when a press/radio war erupted during the mid-1930s "which prevented radio stations from presenting news in any way calculated to damage a newspaper's circulation."[10] The formula was simple: retain the information and journalistic functions for newspapers, and leave entertainment as broadcasting's province. Continuing into the late 1930s, radio news was used as a filler to punctuate the live music that stations and networks turned to during unsponsored periods.[11] Independent station owners at the time feared that AM broadcasting might eventually be entirely taken over by the newspaper industry.[12]

With the introduction of FM radio in 1941 the cycle threatened to repeat itself. This time dual considerations compelled newspaper intervention—the potential for utilizing FM for facsimile transmission, and the fear that FM would quickly render AM stations obsolete. Acting to protect their investments in both fields, newspapers, only months after FM was authorized, had secured 25 percent of the licenses.[13]

With television a commercial reality a scant five years later, the response was much the same. Now publishers felt constrained to protect themselves, their AM, and their FM stations from video competition. Accordingly they gathered in the choice television properties and were well established in the medium by the mid-1950s.

Since that time the total number of newspaper-owned stations has remained fairly stable. And the continued growth of the industry as a whole has made for a gradual percentage decline in crossownership over the years. Particular characteristics of the crossownership situation have changed, however, since the press's first headlong rush into television. As Table 5 shows, the initial movement in the industry was for newspapers to build stations in their local communities. The figures for 1950 reflect the inroads of newspapers into television before the licensing freeze in 1948. With the end of the freeze in 1952, newspapers resumed their sprint for licenses, and within a year they virtually controlled the industry, owning 79 percent of the stations on the air in 1953.[14]

Some saw in this move an effort by newspapers to spread or share their costs, and thus increase their earnings.[15] This would only be the case, however, if newspapers and television were offering the same services. A study by Levin at the time showed, however, that there were no appreciable economies in crossownership. In part, this was because of a marked difference in the services that owners had apportioned to their television stations, and those apportioned to their newspapers.[16]

Between 1953 and 1955, the total number of stations on the air more than doubled, reducing the crossownership share of the industry. The year 1955 was the beginning of the telefilm boom, and local programming virtually came to a halt, with the influx of Hollywood-filmed entertainment fare. After 1955, newspapers not already owning licenses rarely sought them. The flood of newspaper applications for licenses three years before was suddenly reduced to a trickle. The choice VHF properties in the largest markets were already taken by then, in any event. And the Commission, somewhat belatedly, grew increasingly hostile to crossownership. Consequently, newspaper-broadcasting ties stabilized.

Overviewing the patterns in crossownership history, it must be said that newspaper owners, with a watchful eye towards protecting their investments, have been quick to appreciate the potential threats to their security posed by the introduction of new media. For in the introductory phase, the potential uses and functions of a new medium are uncertain, yet to be fixed. Such matters are determined by those first in control. And newspapers have been the first in line to take advantage of new technologies as they have arisen. They everywhere secured radio licenses at the advent of broadcasting, first in AM and then FM, and later moved into television when the video medium seemed destined to supplant its predecessors. As one analyst of broadcasting's early years has noted, such "joint enterprises may restrict the economic development and growth of their electronic subsidiaries, in hopes of shielding their heavier investments in older

TABLE 5

Newspaper Crossownership of Television Stations

	Stations Broad-casting	Newspaper Owned (percent)	Newspaper Owned Stations		
			Nonlocal	Local	Local (percent)
1950	96	43	6	35	85
1955	419	36	27	123	82
1960	533	30	45	116	72
1971	683	24	75	88	54
1974	693	24	91	78	46

Sources: Broadcasting yearbooks and "Statistical Materials on Newspaper-Broadcast Joint Interests," (FCC mimeo material).

media like newspapers."[17] One obvious way to contain the development of a rival medium, of course, is to restrict the character, the range, and the quality of its product, programming. Once newspapers found television less a threat to their survival newspaper ownership of stations leveled off and local crossownership declined considerably. That is, after the introductory phase passed, the only visible newspaper intervention in television has been in realignments which shifted the composition of crossownership. With combined press-broadcasting chains maneuvering to improve their market position by selling some station properties and buying others in larger markets, some local newspaper-television combinations have been separated. In such cases the corporate parent is still a newspaper (or chain of newspapers) but the newly acquired stations have no ties to dailies in their community. Thus while 82 percent of the newspaper-owned stations were locally crossowned in 1955, Table 5 shows that only 54 percent still fitted that description in 1971, even though the total number of stations with newspaper ties remained fairly constant. By 1974, fewer than half of the newspaper owned stations were locally crossowned.

Today, moreover, in over half of the crossownership cases, the ultimate owner is either an out-of-town daily or the corporate headquarters of a nation-wide newspaper group. This is borne out by FCC Commissioner Nicholas Johnson's finding several years ago that in 30 of the top 50 markets at least one station is newspaper-owned and half of these, in turn, are controlled by one of seven national multimedia conglomerates.[18]

FCC CROSSOWNERSHIP POLICY: 1927-72

Governmental concern over common ownership of print and broadcast media dates back to the enactment of the Radio Act of 1927. The issue was first raised by the architect of that bill, and was the subject of further discussion before Senate committees again in the 1930s.[19] During this period when newspaper ownership of radio was increasing rapidly, provoking an intermedia war in the process, it was members of Congress, not the Federal Radio Commission nor the FCC, who became alarmed at the conflict and pressed for ownership diversification. A bill was introduced at the time "to prohibit unified and monopolistic control of broadcasting facilities and printed publications,"[20] but failed to pass.

The FCC at the time scarcely acknowledged the media battles in its midst: no mention of the crossownership issue appeared in licensing proceedings until 1936, and then only in the dissenting opinion of a lone Commissioner.[21] That same year Franklin Roosevelt was elected to a second term, and the elaboration of New Deal economics included a revival of anti-trust issues, with aggressive trust-busting policies favored in the Congress, the Justice Department and the Federal Trade Commission. Under pressure from these prodiversification forces in the Congress, the FCC eventually addressed the crossownership issue, but without modifying the implied policy of nonintervention whether for the purpose of enforcing competition or restraining monopoly. Speaking through its general counsel, the Commission claimed in 1937 that it lacked the authority to prohibit newspaper owners from receiving broadcast licenses solely on ownership grounds.[22] The following year this hands-off policy was reinforced by a staff study. As studies go, this was an exceptionally inconclusive one, the staff reporting that it had no knowledge of the effects of crossownership on the communications flow, if any. Rather than investigate the matter further, it urged upon the Commission a case-by-case approach to crossownership, rather than a general rule.[23]

At the same time the U.S. Court of Appeals for the District of Columbia Circuit, the court with appellate jurisdiction over the actions of all federal administrative agencies, including the FCC, ruled that there existed no statutory basis forbidding newspaper owners from broadcasting.[24] Even so, the FCC paid some deference to diversification currents before the close of the 1930s by twice citing newspaper ownership as a factor in granting radio licenses to independent applicants in comparative hearings.[25]

In 1941 the confluence of a new broadcast band (FM) and a New Deal majority sitting on the Commission produced the FCC's first

broad-scaled attack on crossownership and its first attempt at crossownership rulemaking. Newspaper owners were aggressively seeking
licenses for the FM spectrum which the Commission had just then begun to authorize. At the urging of President Roosevelt and in the ab-
sense of Congressional action, the Commission launched its newspaper
investigation in the spring of 1941 and simultaneously halted further
licensing, construction and transfer of newspaper-owned FM stations.[26] The investigative hearing, called for the purpose of establishing rules on crossownership, was variously interpreted as an
attempt to stimulate Congressional activity,[27] and as retaliation
against the nation's newspaper publishers for their solid support for
the Republican candidate in the 1940 presidential election.[28]

Consequently the publishers balked when called upon to testify.
Unable to summon its star witnesses, the FCC went to the D.C. Circuit Court, where the Commission's investigative authority was affirmed. But the move seemed to backfire when the court warned the
FCC that it lacked the power to ban crossownership outright.[29] In
effect this granted the publishers a reprieve from any diversification
action by the Commission as the Congress, then consumed with war-
time legislation, stood mute. Thus citing the "grave legal and policy
questions" involved, the FCC closed its hearings three years after
their inception without adopting any rules. At the same time the Commission served warning to the publishers by reaffirming its intent to
prevent "concentration of control" on a case-by-case basis.[30]

In the ensuing years the Commission showed the first signs of
a serious intent to hold down the extent of crossownership, refusing
radio licenses in 1946 to nine newspaper applicants.[31] Support for
diversification was stirred a year later when the Commission on Freedom of the Press warned of an increase in local news monopolies.[32]
Their report drew attention to situations in 100 communities where
the sole newspaper owned the only broadcast station.

Yet the pattern of licensing decisions over time reflected the
Commission's desire not to appear to have created a de facto ban
against crossownership under the guise of the case-by-case approach.
Thus the Commission's radio license awards were dividing evenly in
the relevant contests, half favoring newspaper owners, half disfavor-
ing.[33]

Then in 1950 and in 1951, the D.C. Circuit Court in the Mansfield Journal and Scripps-Howard cases at last affirmed the FCC's
authority to license on the basis of crossownership.[34]

On the heels of these new court rulings and with the freeze in
television licensing about to be lifted, newspaper interests grew
alarmed that their intent to move into radio would be frustrated. Thus
they quickly sought and found an attentive ear in Congress.

During the Roosevelt years especially, Congress had been a province of propluralism sentiments. But the Congress in session in 1952 was the same one that amended the Communications Act to limit FCC intervention in station sales.* That year it proposed a Newspaper Amendment to prevent FCC discrimination against publishers seeking broadcast licenses. With congressional passage certain, the FCC realized it would be better off with its flexible case approach than Congress' restrictive rule. The FCC's chairman was therefore sent to Congress to pledge that the Commission had not and did not intend to discriminate against newspaper owners.[35] The amendment was subsequently dropped.

The FCC complied with its contract with the Congress during the first full year following the freeze. And newspaper ownership of television stations jumped from 45 percent to 79 percent of stations on the air.

But then in 1954, newspaper publishers were passed over in two television license contests and in each case newspaper ownership, for the first time, was cited as the chief factor in the decision.[36]

Judicial support for such action, and for diversification generally, grew steadily after 1955 in a series of court decisions by the D.C. Circuit Court and the U.S. Supreme Court.[37] Buttressed by the prevailing judicial posture, the FCC, by the end of the decade, had evolved a policy which in general preferred nonnewspaper owners to equally qualified newspaper applicants. Throughout this period the FCC reaffirmed its belief that nonnewspaper owners would be more likely to promote diversity of expression.

Commission doctrine still fell short of an absolute ban on newspaper ownership in newly issued licenses. And it did nothing to reverse the actions of earlier years which had given publishers first, virtual control of television, but by now only a major share in the industry.

Legislation was introduced in Congress in 1960 "to prohibit the concentration of control of a substantial portion of the television and radio broadcasting facilities and . . . of the news publications in any section of the country."[38] Had it passed, the legislation would have given the government grounds to compel some newspapers and newspaper chains to sell their broadcasting stations.

Except for a brief interlude during the chairmanship of Newton Minow (which will be considered later), the Commission continued to oppose crossownership on the grounds that it eroded business pluralism. This was made plain in the Commission's 1965 Policy Statement

---

*See Chapter 2, pp. 20-21.

on Comparative Broadcast Hearings wherein "a maximum diffusion of
control of the media of mass communications" was cited as one of the
"two primary objectives" to be satisfied in licensing proceedings.[39]
The statement identified local crossownership as the most salient target of the FCC's diversification objective.

Commission policy during the 1960s, geared as it was to initial
licensing proceedings, did nothing to alleviate media ownership patterns created during television's early years. In the 1960s, newspaper-TV crossownership had stabilized, advancing or receding only
slightly from year to year. But there remained by the end of that decade over 150 newspaper-owned video broadcasters securely protected
by a pluralism which applied only to initial license grants. These TV-
press combinations could survive and even increase in number so long
as license renewal and sales proceedings were not subject to the diversification policy.

In 1968 the Justice Department embarked upon a series of ini-
tiatives calculated to prod the FCC into adopting more restrictive
rules on newspaper-broadcasting crossownership. In May of that year
the Department advised the Commission that it might contest FCC
decisions perpetuating or authorizing further newspaper control over
broadcast stations.[40] In the absence of an affirmative response, in
August the Department moved to block the sale of a Beaumont, Texas
TV station to the owner of the only daily newspapers in that city.[41]
Simultaneously, the Department issued a memorandum to the FCC
urging the outright prohibition of local newspaper-broadcasting crossownership.[42]

At the time of the memorandum the Commission had barely
commenced rule-making hearings on a milder variant of the crossownership controversy, that involving TV-radio ownership combinations.[43] The memorandum suggested that these hearings be expanded
to include the weightier matter of local newspaper crossownership.
The memorandum's second bit of advice was more explosive: it urged
the FCC to enforce its professed pluralism by denying license renewals to local newspaper-owned stations.

With the FCC still mute, the Justice Department in December
filed what it described as its first antitrust suit contesting the merger
of newspaper and television interests, against a Rockford, Illinois
media combination.[44] The same day of the filing, the owners announced they would sell the station rather than contest the suit. Brist-
ling with success in both this and the earlier Beaumont action (where
the intended merger had been called off by the owners), the Department called upon the FCC in January 1969 to take action on its own
to break up the "mass media communications monopoly" of the
McCracken family in Cheyenne, Wyoming.[45]

Woefully upstaged at this point, the Commission displayed in a matter of weeks a newly discovered determination in the whole matter and issued a jolt of its own. It took from the Boston Herald-Traveler a TV station it had operated for more than a decade, WHDH, and awarded the license instead to a rival Boston business group free of other media interests.[46] The decision was without precedent. In the course of literally thousands of TV licenses up for renewal over the FCC's history, this was the only instance of revocation or denial on pluralist grounds.* Since it happened to involve a broadcast property estimated to be worth $50 million on the open market, the decision left the broadcast industry aghast.[47] Statements from Commissioner Nicholas Johnson at the time, offering WHDH as a precedent for future local takeovers in other crossownership cases up for renewal, provided the industry no reassurance.[48]

In retaliation against the FCC's rare indiscretion of exercising its statutory authority in license renewal proceedings, the industry went to the Senate. There it secured the introduction of a bill which, if passed, would have virtually eliminated the public's historic right to simply challenge station owners at license renewal time.[49] Until the WHDH decision, it had been a paper right in any event.

The FCC's initiative in the WHDH case thus presaged a new Commission activism and seriousness of purpose in behalf of business pluralism, at least with regard to local newspaper crossownership.

Support for the Commission's lead followed shortly from other government quarters. In November, 1969, the Mass Media Task Force recommended to the President's Commission on the Causes and Prevention of Violence that "except in cases of above average performance, license renewals by television stations affiliated with a newspaper in the same community should be granted only on the condition that the station or newspaper is sold within the next three years."[50] And the following January a bill was introduced into the Senate to prohibit owners of daily newspapers from acquiring or continuing to own radio or TV stations in the community served by the newspaper.[51]

---

*The WHDH hearings were the longest on record, having commenced in 1954 when six original applicants sought the channel. Thus the winner in the case contends that the original license grant proceeding was never completed, meaning that it prevailed over the newspaper controlled operator of the station in what was actually an initial licensing comparative hearing and not, strictly speaking, a renewal action. The Commission staff, nevertheless, has persistently construed the matter as a renewal case.

Meanwhile the FCC showed further signs of an impending crack-down. It first set for hearing the license renewal applications of news-paper-owned stations in San Francisco and Minneapolis.[52] Then it denied the sale of a Wichita station of a newspaper chain owner.[53]

Finally in March, 1970, the Commission announced an extension of its crossownership rule-making proceedings, having decided to follow the advice of the Justice Department offered nearly two years earlier. Rules were proposed barring local newspaper-broadcasting crossownership altogether, requiring divestiture to end existing in-stances of local newspaper-television crossownership.[54]

THE INDUSTRY STANCE: BOTH SIDES NOW

In the hearings that followed the broadcasting industry took a position which both commended business pluralism and rebuked it. It did this with two separate studies within the same hearing.

First the industry acknowledged the FCC's deep concern for a pluralist ownership structure, and attempted to persuade the Com-mission that pluralism in ownership had never been more pronounced than at present. The industry seemed to be telling the Commission that it understood and accepted the premise of pluralism, but that the Commission's fears for pluralistic ownership were ill-founded, per-haps even misplaced. This position was argued, however, not by fo-cusing on the ownership patterns in broadcasting, but precisely the opposite, by distracting the Commission's attention away from its statutorily designated area of regulation and expertise. The industry directed the FCC to look at the pluralism broadcasters claimed ex-isted over all types of media: print and broadcast, national and local, daily and weekly, low circulation and high circulation. Here, it was alleged, there was a vast multiplicity of separate ownerships, and by implication, the FCC's concern over broadcasting's state of affairs was made to seem rather parochial.

But the defense did not rest there. In a separate filing, the in-dustry, joined this time by newspaper publishers and their chain owners, instructed the Commission to recall what its pluralist man-date had been designed to achieve. Surely not pluralism for its own sake, they reminded, but rather a greater diversity of free expres-sion in the communications flow. Was not the FCC's goal an expanded, more open information structure? This, they claimed, was in fact the contribution made by newspapers whenever they had been allowed to own television stations. Thus to prohibit crossownership would be self-defeating.

On the one hand the Commission was being extolled for its interest in pluralism. On the other, the FCC was told not to accept the assumptions about pluralism's efficacy so uncritically, but to return to
its first principles. These arguments kept the FCC pondering its proposed rule to ban local crossownership for five years, in the process
stalling the Commission's earlier accelerating interest in enforcing
pluralist policies in this area. Accordingly these arguments deserve
some attention, the more so because of their gratuitous pose.

The industry's pro-pluralist stance was embodied in the Seiden
report, commissioned by the National Association of Broadcasters
and filed in early 1971 on their behalf in the FCC's crossownership
hearings.[55] The Seiden report advances a pluralist image of a multitude of media fiercely competing for the attention of local audiences.
Television stations, radio stations, daily newspapers, weekly newspapers and magazines are all seen as equal media voices providing a
proliferation of messages to the information consumer that borders
on saturation. In this context, it is argued, the instance of common
ownership of daily newspapers and a television station in the same
community assumes minor significance. This is because the number
of alternative media and even the number of alternative media owners
is so extensive.

The report consists simply of an enumeration of all the monthly
and weekly magazines which have sales of 100 or more per issue in a
given metropolitan area; plus the newspapers and broadcast stations
which penetrate 100 or more homes in the designated metropolis. The
result is a pluralism score for each media market in the United States.
This accounting shows, for example, that in New York City, the nation's largest media market, there are no fewer than 610 different
media voices operated by 434 separate owners. In view of such pluralism, can ownership of the area's largest circulation daily newspaper in common with one of its VHF television stations seriously be
alleged to pose a threat to diversity of expression? Looking at smaller
markets, such findings, on a less impressive scale, are claimed to
merit the same conclusion. Bellingham, Washington, ranked number
190 in market size with only 24,000 homes, shows 55 different media
voices operated by 46 separate owners. The effect of such media
counts is clearly to substantiate claims that even outside of large
metropolitan areas Americans are offered a large number of alternative media sources to choose from. The broadcasting trade press
concludes that this apparent pluralism of media voices casts a different perspective on communications problems: "the problem in communications is not one of concentration of control of the media; it is
the problem of the communicator in making his voice heard above the
din created by his competitors."[56]

More by calculation than coincidence, these market-by-market counts add more confusion than clarity to the concept of pluralism in question. The circulation patterns for such national magazines as Family Circle, Field and Stream, Forbes and Fortune (each included in the Seiden count) may provide some interesting data on periodical pluralism. But neither magazines nor newspapers bear any resemblance to broadcasting on the point critical to the pluralist controversy. Broadcasting alone is a scarce national resource, owned in theory by the people, and therefore accountable through the regulatory mechanisms of the state to facilitate free expression and the interests of broad and diverse publics. The other media bear no such legal status. Their owners may freely pursue their own exclusive interests, whether profit, propaganda or a combination of both, without assaulting any legal guarantees to their subscribers. On the other hand, broadcasting is the sole mass medium from which the public has a right to expect an open and diverse communications flow, which, in turn, the state is empowered to facilitate.

Such fundamental distinctions were ignored by Seiden's enumeration method of assessing pluralism. Yet one can appreciate the evident appeal of his approach to the broadcast industry if in fact television stations are no more accessible to diverse local expression than are the national headquarters of mass circulation monthly magazines.

Quite apart from such distinctions, the Seiden technique artificially inflates the evidence in favor of gross media pluralism, even within the terms of his all-inclusive definition. Not only are national periodicals added, but television satellite stations, which simply duplicate the programming of their parent outlets, are separately enumerated, as are broadcast stations located over 100 miles from a given locale.

The results of such an approach can perhaps best be demonstrated by noting the Seiden figures for Cheyenne, Wyoming, a market well-known for its mass-media monopoly. In Cheyenne, at the time of the Seiden survey, the McCracken family owned the only daily newspapers, the only local television station, the only full-time local AM station, and the only cable television system serving the area. A look at the Seiden figures, however, gives no clue whatsoever to the distinctive situation in Cheyenne, least of all a hint of a communications monopoly. The report shows 114 different media reaching the community, operated by 96 separate owners, including 45 voices originating in the market, controlled by 38 separate owners. To the unknowledgeable, Cheyenne would appear to have a highly respectable media structure. But surely a technique which fails to even identify one of the most flagrant instances of monopoly cannot be said to furnish a serious perspective within which to view the pluralism problem.

Indeed, the whole method of undifferentiated nose-counting is viewed skeptically even within certain industry circles. Several months before the Seiden report was made public, a group of news-paper-television owners filed a petition with the FCC. Their appeal attacked the presumption "that meaningful diversity of voices flows from mere numbers. . . . We believe the presumption has become terribly rustic and unsophisticated," the petitioners stated, claiming that it "had become mature dogma which many regulators and critics accepted and continue to accept on faith."[57] Ironically, only months later the owners' trade association filed the costly Seiden brief, par-roting the same presumption these owners earlier disparaged.

For its own part, the Commission remains unimpressed that weekly newspapers, national periodicals and distant broadcasting sta-tions are adequate safeguards against monopolistic controls over lo-cal expression.[58] It may, after all, have been just such reasoning which prompted the American Newspaper Publisher's Association (ANPA) to file briefs and studies separately from the NAB in the crossownership hearings.

Given the difficulties of this pluralist line of defense, it is no surprise to find a complementary argument advanced in support of crossownership to the effect that newspaper-owned stations excel in the crucial performance areas—news and public affairs information. The basis for this assumption is engaging. Newspapers, most would agree, are in the business of news and not, as with broadcasters, primarily entertainment. Journalists make up the core of their staff, not announcers and talent artists. Presumably the journalistic heri-tage of the oldest mass medium would naturally be infused, where common ownership made it possible, into the newest mass medium. Thus it is contended that the product goals, the journalistic values, and the professionalism of the news enterprise could, through the en-tanglement of their economies, be transferred from the daily news-paper to the television broadcaster.

In the course of repetition this journalistic legacy theory some-times even found adherents on the FCC, most notably in the person of a Commission chairman who at the time was widely mistaken for a reformer and unmerciful critic of the industry.[59] During Congres-sional hearings before a subcommittee investigating media monopolies in 1963, the then FCC Chairman Newton Minow testified

> Though I could not document this, it is my personal im-
> pression that some of our broadcast licenses which are
> affiliated with newspapers and periodicals are among
> those broadcasters most serious about service to the pub-
> lic interest. Some of them who have come to broadcasting
> from a tradition of journalism rather than entertainment,

have set high standards of independence from advertisers,
of emphasis upon informative broadcasting with extensive
news staffs, and upon dedication to meeting community
needs and advancing community needs and advancing community projects.60

So impressed, in fact, was Minow with his own unverified assumptions, that he proceeded to advocate giving television stations
to each and every daily newspaper owner in the top 25 television markets, a feat which if performed at the time would have resulted in
three-quarters of all such stations being controlled by the local
press.61 Earlier statements by Minow in the same vein were seized
upon by executives of the American Newspaper Publisher's Association and used repeatedly to their advantage in later hearings.62

It should be noted that assumptions about the blessings bestowed
upon broadcasters by their newspaper owners suffer from an inadequate appreciation of the economic history underlying such combinations. We have already seen how, in the course of successive intro-
ductions of new communications technologies, newspaper owners have
consistently dominated each new medium's managerial structure during its developmental and potentially most innovative stage. Newspaper intervention, coupled with the consequent distinct division of
specializations among the media, suggests that press supervision has
functioned more to control competition in the information flow than to
foster it.

Nonetheless, the presumption of goal and value transference
from newspapers to their economic partners in broadcasting persists.
And attempts have been made in recent confrontations to substantiate
this claim.

One noteworthy instance occurred during Senate hearings in
1969 on the bill which the broadcasting industry proposed immediately
following the FCC's historic denial of a license to the newspaper
owned WHDH-TV. Appearing in support of the bill, executives of the
ANPA testified that "as a group, newspaper-owned broadcasting stations over the years have compiled a record of distinguished achievement and meritorious service to the public."63 As evidence they
produced a study, prepared under their direction by a journalism professor client, which analyzed ownership characteristics of recipients
of the University of Georgia's annual Peabody Radio and Television
Awards. The awards which "honor excellence in programming . . .
are highly regarded in the broadcasting industry."64 The conclusions
of the study were as follows:

Thus, 31 out of 124 possible awards went to newspaper affiliated stations for a ratio of 1:4. Expressed another way,

one-fourth of all awards to individual stations went to
newspaper-oriented or affiliated stations.

Since only 10.2% of the broadcasting stations in the
United States are newspaper oriented or related, the
ratio of those winning Peabody Awards is 2 1/2 times
greater for newspaper-oriented stations than for all individual stations.

This, in our estimation, is a significant differential
and based upon our background with the Peabody Awards
clearly indicates the feeling of the Board over the past
28 years that newspaper-oriented stations have demonstrated distinguished achievement and meritorious public service in ways substantially greater than the average
non-newspaper oriented broadcasting station.[65]

Although radio and television stations are mixed in the study's
count, and book and periodical publishers are included as ''newspaper-
oriented'' and ''related'' owners, the Peabody data provide some empirical indication that newspaper-owned broadcast outlets excel in
journalistic performance. Moreover, the veracity of such a conclusion would be enhanced if it originated from the offices of an impartial
university board of journalism critics. However, what went unre-
marked at these hearings was that at the time of this study, individuals
such as NAB executives, newspaper editors of newspapers owning co-
located television stations, and celebrated advocates of crossownership such as Newton Minow, were among the members of the Peabody
Awards Committee.

The FCC hearings on proposed rules to ban local crossownership provoked further efforts among the industry's lobbyists to make
a compellingly affirmative showing of the journalistic legacies in local crossownership. The result is another edition in the long series
of industry-financed studies.[66] Basically it attempts to showcase the
performance claims of crossowners through substantiating field research.

The NAB's offer of $50,000 was enough to procure Professor
George H. Litwin of the Harvard Business School and William H.
Wroth of Intermedia Systems Corporation to coin this study which
has been branded ''a deal from a stacked deck'' by one careful
critic.[67] The FCC was even less charitable in its review: ''we
found it of little value.''[68] In their preface the authors claim un-
blinkingly to have undertaken their research ''not to argue or defend
any specific position regarding common ownership,'' but rather ''to
provide an organized and systematic research basis for the development of public policy regarding common ownership.''[69] Notwithstand-
ing such claims, the study is a compendium of the principal arguments

in defense of crossownership of all types (TV-radio and newspaper-broadcasting—the authors prefer the term common ownership). Eschewing a ritualistic march through the data the authors conclude on page two that "the findings strongly suggest that the establishment of across-the-board rules limiting common ownership would be detrimental to the public interest in a majority of cases."[70]

The expense of the study, the prior credentials of its authors, its emphasis on field research, and because, as the study itself asserts, similar research has earlier moved the FCC to modify its policies—for these reasons the Litwin study must be regarded as a major statement in defense of crossownership interests, but not as an empirical substantiation of them.

The research reported is an intensive study of six television markets, one with high local crossownership and one with low local crossownership in each of three different market size categories (large, medium and small). All of the data came from interviews with media owners and managers and local elites. Thus the data provide an unusually purposive sample of opinions on the impact of crossownership and practical differences between crossowned and other media.

Before turning to the study's findings, an appreciation of its shortcomings is in order. In an elaborate critique of the Litwin study, Barnett found that "the research methods employed by Litwin were so biased in favor of common ownership, and the premises so arbitrarily contrived to the same end, as to vitiate all the findings and conclusions opposed to diversification."[71] The FCC, a body not heretofore known for its expertise in research design, itself observed that "the sample selection was completely biased in favor of common owners."[72] Scattered throughout the Commission's evaluation are references to Litwin's inappropriate data, unwarranted assumptions, uncontrolled analysis, and tendencies to make conclusions based on differences shown to be not statistically significant.[73] The Commission's verdict was that the most important conclusions of the study were simply invalid.

Inasmuch as the study's objectivity has elsewhere been challenged and its findings dismissed, one might question the need to contend with the Litwin study at all. Yet, even if it fails to substantiate industry claims about crossownership (a fact acknowledged implicitly by the NAB's hasty commissioning of more studies in the wake of the Litwin critiques), it is nevertheless a revealing statement of those claims. Even if it fails to provide the "research basis for the development of public policy," it at least sets forth the crossowners' position regarding their distinctive contribution to the broadcast communications structure.

Briefly stated, the arguments favoring local crossownership are advanced by Litwin as follows:

Commonly owned media have larger news staffs, do more
news programming, and are less dependent on the wire
services and networks for news than singly-owned media.

. . .

In singly-owned media, profitability is the first-ranked
goal, followed by news coverage. . . . In commonly-owned
media, news coverage is the first-ranked goal, followed
by public service . . . (profitability is ranked fifth).

. . .

Commonly-owned media are perceived by business
and community leaders as providing greater validity and
depth of news coverage, better quality programs, more
public service, and as having higher business ethics.[74]

Litwin's comparisons translate into superlative ratings for
crossownership. The drift of his arguments is clear: crossowned
media are more dedicated to news and public service values and to
the professional journalistic dimension of the mass media generally
than are broadcast stations without newspaper ties. In short, locally
crossowned television stations have not only inherited the journalistic
legacy of their investors, they have improved upon it, and, in the
course of the transaction, acquired an uncharacteristic distaste for
profits. This is hardly a surprising argument given the predilections
and vulnerabilities of the FCC in the matter.

Like the opinions of Minow considered previously, the Litwin
statements are based on perceptions, and not actual performances.
Nonetheless, the arguments raised here, and earlier in the Peabody
Survey, and still earlier by other defenders of crossownership, have,
as we have seen, disputed the merits of a strict adherence to business
pluralism. Fortunately, they serve to cast the crossownership issue
into conveniently researchable terms: Does local crossownership facilitate diverse expression and increased attention to public affairs
as evidenced by a different shape in the communications flow from
crossowned television stations?

So framed, the question speaks unequivocally to the crossownership issue as defined in an emergent consensus of commissioners,
crossowners and those self-cast as public advocates in this matter.
The FCC has called for objective, factual data on program service.[75]
Industry watchdogs have gone on record saying that performance data
"represent(s) the single most decisive element in evaluating the effects" of the rule "on the public interest."[76] And an assemblage of
crossowners has found at "the heart of the matter, the quality of
service to the public" and offered as a criterion a

> mass communications media which . . . will . . . con-
> stantly impart to the public a greater and greater por-
> tion of knowledge and culture which it needs if . . . the
> complex questions confronting our society are to be
> resolved through the democratic process.[77]

Other major parties to the policy issue have similarly attested to the
decisiveness of performance evidence in resolving the fate of crossownership.[78]

The chief evidence introduced at the hearings to support the
FCC's strict pluralism and to ban crossownership was arranged by
some well-established public advocacy groups, including the National
Citizens Committee for Broadcasting, and authored by Harvey J.
Levin.[79] Levin is an economist who five years before collaborated
with an NAB-funded research team to oppose pluralism as it applied
to group ownership.

To evidence the advantage of pluralism, or at least to evidence
the lack of any attendant disadvantage, Levin used a concept of communications diversity favored by the industry. This concept has been
carefully cultivated under industry auspices in earlier published books,
generally laudatory about the existing nature of television communications.[80] The concept equates diversity in communications with the
number of viewer choices among program categories over all stations
in a market at any moment in time. In other words, the measure indicates the variety available to the viewer by flicking the dial. But the
industry's and Levin's idea of variety, which each is prepared to call
diversity, reveals more about the constricted imagination that passes
for policy analysis in regulatory deliberations than it does about the
communications flow. For example, if in a three station market one
station broadcasts all cartoons, another all quiz shows, and the third
all situation comedies (these are "diverse" categories as used by
Levin), then using his measure diversity in communications would be
maximized. This is because the maximum number of viewer choices—
three—would exist at any given time. Comparing television markets
with crossowned stations and markets without them, Levin concluded
that "the non-newspaper stations in our study have contributed significantly more to diversity and choice."[81]

The more such analyses are represented as briefs in behalf of
audience interests, and as a basis upon which to condition regulatory
policy, the easier it is to appreciate how thoroughly conditioned so-
called public advocates are to finding the present broadcast system
normal, and its business readily acceptable.

More germane to the relationship between business pluralism
and the communications flow is a preliminary analysis of prime-time

programming which Levin attached to his research on diversity. He surveyed allocations of air time to public affairs, news, and locally produced programs during evening hours, and concluded that "in none of these programs types . . . do the newspaper owners systematically outclass the non-newspaper licensees."[82] Unfortunately his efforts here are marred by tendencies in his discussion to contradict the data printed in his tables.*

Our own data from the national programming survey reported on in the previous chapter is shown in Table 6. Because newspaper owners sought and received television licenses in the first wave of Commission allocations, crossowned stations are overwhelmingly VHF broadcasters and network affiliates. Consequently the table compares only VHF, network-affiliated TV stations. Crossownership is broken down into two categories: stations with ownership ties to newspapers in their own community, and a separate category including all stations with newspaper owners, whether local or nonlocal.

Yet an examination of the figures in Table 6 shows that these ownership distinctions, however drawn, are meaningless with respect to a willingness to service public information and expression. The shape of the communications flow is remarkably consistent from station to station whether owned by a local paper, owned by newspapers located elsewhere, or free of newspaper ownership altogether. The shape of the communications flow among these ownership types is likewise consistent with that for the industry as a whole, considered in the previous chapter. Even within the small component of air time allocated to information, in moving from general news, to public affairs, to local expression formats, the flow is rapidly attenuated to the point where it becomes imperceptible, or more accurately, nonexistent.

In the case of each program category, the statistical tests indicate that there are no true differences among the means for each ownership type.

---

*As one example, he claims that among network affiliates in multi-station markets, newspaper-owned VHF stations offer statistically significantly fewer news and public affairs programs in the prime time period than those nonnewspaper-owned. A look at his table reveals very marginal differences which are indeed significant statistically, but in the opposite direction. Newspaper owned stations show a slight statistical advantage in news and public affairs programming. And for whatever reason, even in the table the data are reported in a form such that they appear to support Levin's argument. Elementary arithmetic, however, shows the data to be contradictory to his description.

TABLE 6

Television Informational Programming of Newspaper-
and Nonnewspaper-owned VHF Network Affiliates
in the Top 75 Markets, 1970

| | Mean Percent of Broadcast Week | | |
| | Newspaper-owned Stations | | Stations Not |
Programming	Local N = 21	Local and Non-Local N = 26	Newspaper-owned N = 61
Local public affairs	0.9	0.7	1.0
All public affairs	2.2	2.0	2.3
Local news	7.5	7.2	7.2
All news	12.2	11.7	10.7

Source: Compiled by the author.

Only in the total news category do the observed means differ by
more than a fraction of a percentage point. While a difference here of
1.5 percent is short of substantive import, it even proves to be an artifact of certain characteristics of ownership history. A disproportionate majority of newspaper-owned stations are affiliated with the CBS
or NBC networks, in contrast with a disproportionate number of ABC
affiliates among nonnewspaper stations. This is because the earlier
entry of newspapers into the video medium afforded them a choice of
networks. From the beginning, CBS and NBC were the favored affiliations, and not until 1954 was ABC able to offer its affiliates a regular
daytime schedule of programs. By that time newspaper owners had
79 percent of the licenses and were well entrenched. Thus the earliest
broadcasters, and this meant most of the newspaper owners, looked
first to CBS and NBC for affiliation. And in the case of CBS, press-
owned stations were virtually guaranteed a hookup. Until at least
1958, CBS was on record as favoring newspaper-owned stations for
network affiliation.[83] Even now ABC offers less news programming
to its affiliates, on a daily basis, than do NBC and CBS. Thus this
minor difference in total news (which is the sum of network and local
news) is attributable simply to network affiliation differences between
the ownership types.

Accordingly, when network affiliation was controlled for within each ownership type, the programming differences in total news between ownership types proved to be only fractional, as in the other programming classes. And applying statistical tests here showed, once again, that in all of the programming classes, each of the ownership types shares the same true mean.

The implications from these data are twofold. First they show that adherents of a strict pluralist policy, who would have that policy extended to eliminate crossownership, can expect nothing in the way of a change in the communications flow or the information structure through pluralism so conceived. Second, claims about an enviable and unmistakable legacy of journalistic values and public-service responsibilities for newspaper-owned television stations are not substantiated when one looks at their actual allocations of broadcast time to informational programs. Nor are they substantiated by comparing crossowned stations with the rest of the television industry. Accordingly, those who maintain that the original goals anticipated by pluralism are best achieved through policies which permit newspaper crossownership speak of illusory hopes, not informed appraisals. For neither a business pluralism which bans crossownership, nor a policy which tolerates newspaper-television monopolies with a belief in their journalistic heritage, can be expected to alter noticeably the prevailing practices and priorities in electronic communications. These, it seems, are homogeneous with respect to the distinctions in ownership considered here.

But if crossownership is without public dividends, private dividends are another matter. Economic integration of broadcasting and the daily printed press assures some control and protection against intermedia competition for advertising and audiences. With this in mind, publishers moved successively into each new broadcast medium, first AM, then FM, and finally television. In the early days of radio, publisher-broadcasters were more conspicuous in their efforts to shape the new medium along lines that prevented threats to printed press stability. The passage of time may have improved the sophistication of their efforts, but not the consequences. It is surely more than coincidental that today daily newspapers receive most of their revenue from local advertising, while national advertising comprises the bulk of television revenue; that the daily press is still primarily a news organ, while TV is essentially an entertainment medium; that the daily papers rely heavily on local news, while TV does very little local programming whatsoever.

When it was still a novelty, video broadcasting technology was established with local service in mind. As quickly as the vision of plural local information sources threatened to become a possibility, the reality of a few corporate enterprises intervened. Controlling the source of threats to competition was apparently worth the investment since, as we have seen, there were no cost savings or economies of scale obtained through crossownership.

As regards the journalistic legacy, it is argued that newspaper ownership of television should carry with it the heritage of hard-won struggles for speech and press freedom, the ideals of journalists like Locke, Milton, Mill, Zenger and Paine. The commitment to free expression and investigative journalism should be transferred to broadcasting by publishers to a degree not possible by owners lacking the inheritance of the Fourth Estate.

But the more appropriate saints, it turns out, are James M. Cox, Samuel I. Newhouse and E. W. Scripps, names perhaps equally familiar to students of business as to those of journalism, and all press lords whose corporate heirs have reproduced in broadcasting the monopolistic patterns their mentors forged in newspaper publishing. The journalistic legacy theory is thus a neat but unsuccessful attempt to rationalize and legitimate economic self-interest in terms of the public interest. The more probable legacy of crossownership has been a formative restraint on the information potential of television, and complicity in pursuits which have kept video technology from developing into new, independent outlets of local expression.

When the FCC finally handed down a decision in the crossownership rulemaking five years after its inception, only one commissioner remained who had sat on the Commission at the time the rule was proposed. Over the years Commission attorneys had used the pending rulemaking to defend against appeals of Commission license renewals. When license challengers appealed in court that incumbent licensees monopolized local media, Commission attorneys would argue that the FCC preferred not to deal with these important concentration of ownership issues on a case-by-case basis but rather in its general rulemaking proceedings, such as the crossownership proceeding which was always alleged to be nearly completed. It was the Department of Justice, whose actions five years earlier had instigated the rulemaking, which hit upon a strategy that provoked the Commission into issuing a rule in early 1975. During the previous year the Justice Department filed petitions to deny against the license renewals of eight crossowned stations. Finally the Commission acted by voting a ban against prospective local crossownership and ordering the breakup of newspaper-television combinations in seven communities, six of them outside the 100 largest markets. These seven were distinguished from the total of 78 locally crossowned stations by each being the only station in their respective towns, and by owning the only daily newspaper in town. In the decision the FCC verbally reaffirmed its belief that it is "unrealistic to expect true diversity from a commonly owned station-newspaper combination."[84] But the

Commissioners seemed once again content to merely proclaim their
faith in pluralism, refusing to insist on an across-the-board enforcement of this finding. In so doing they noted that the divestitures en-
tailed might prove economically disruptive to the business of broadcasting.

AFTERMATH

     For more than a quarter century of television broadcasting, the
public interest in such communication has been entrusted to the presumed efficacy of a business pluralism. Yet the gap between regulatory rhetoric and television's real business structure has been pronounced. Consequently this gap has shaped controversy and conflict
over regulatory actions. Those bearing a reformist mantle within the
Commission; activists in the Justice Department and the judiciary who
find any disjuncture between policy and practice professionally disturbing; and critics outside of government who despoil the bread-and-
circuses use of the medium and wonder about the fate of free expression and social intelligence in such an atmosphere—these and others
have beseeched the Commission to not merely profess its faith in
business pluralism, but to live by it.
     With the conflict so defined, there appears to have been little
critical space remaining to ponder whether this received wisdom is
in fact so wise. In thinking about the airwaves as a public resource,
about communications inequalities and what a democratic information
structure might look like, to stake out business monopoly and business
pluralism as the static and progressive alternatives, respectively,
seems a discredit to the meaning of alternatives.
     But so too should the reforms which would recapture business
pluralism be discredited. It is an imaginary ideal. Manipulating subtle variations in the business structure, along the lines carved out in
regulatory postulates, might be useful chiefly to demonstrate their
ineptitude. For in the interstices of the station chains, both networks
and groups, where local ownership survives as something more than
an ideal, there is no evidence of an enlightened interest in servicing
free expression and informational needs, any more than that found
among other broadcast owners. As to the willingness to provide forum
opportunities, there seems to be no differentiation within the business
ownership structure. This is also the case with respect to stations
free of the bane of newspaper interests. Audiences of the latter still
await compelling evidence of their often touted journalistic legacy.
     The paradox of all this, however, is that the matter of owner-
ship is not without consequences. The legal foundations of the medium

protect the owners' right to free expression and make the public's comparable right a matter of the owners' privilege—to dispense freely or not at all, as they see fit. This much is acknowledged by all parties. And the owners themselves might admit to even more: in the words of Niles Trammell, president of NBC at the dawn of the television era, "business control means complete control."[85] Accordingly, the nature of business control must have some effect on the nature of the communications flow.

Yet as television celebrated its twenty-fifth anniversary, there were signs that the Commission was changing its views as to the importance of the question, Who owns the airwaves?

NOTES

1. Ben Bagdikian, The Information Machines (New York: Harper & Row, 1971), p. 127.

2. Sydney W. Head, Broadcasting in America, 2nd. Ed. (Boston: Houghton Mifflin, 1972), p. 198.

3. Jerome H. Heckman, "Diversification of Control of the Media of Mass Communication—Policy or Fallacy?", Georgetown Law Journal 42 (1954): 379.

4. Harvey J. Levin, Broadcast Regulation and Joint Ownership of Media (New York: New York University Press, 1960), p. 53.

5. Ibid., p. 47.

6. Christopher H. Sterling, "Newspaper Ownership of Broadcast Stations, 1920-68," Journalism Quarterly 46, no. 2 (1969): 229.

7. Alfred M. Goldsmith and Austin C. Lescarboura, This Thing Called Broadcasting (New York: Henry Holt, 1930), p. 71; quoted by Sterling.

8. Karl A. Bickel, New Empires: The Newspaper and the Radio (Philadelphia: Lippincott, 1930), p. 61.

9. Sterling, op. cit., p. 229.

10. Ibid.

11. Erik Barnouw, The Golden Web (New York: Oxford University Press, 1968), pp. 74-83.

12. Howard N. Gilbert, "Newspaper-Radio Joint Ownership: Unblest Be the Tie That Binds," Yale Law Journal 59 (1950): 1342.

13. Sterling, op. cit., p. 231.

14. Levin, Broadcast Regulation. . .Media, p. 53.

15. See, for example "Diversification and the Public Interest," Yale Law Journal 66 (1957): 365, 367-68; these claims are reviewed in Levin, ibid., pp. 89-90.

16. Harvey J. Levin, "Economies in Cross-Channel Affiliation of Media," Journalism Quarterly 31 (1954): 167; and Levin, Broadcast Regulation. . .Media, p. 92.

17. Levin, Broadcast Regulation. . .Media, p. 75.

18. Nicholas Johnson, "The Media Barons and the Public Interest,"Atlantic (June 1968), p. 48.

19. Testimony of Newton N. Minow before the Antitrust Subcommittee of the House Judiciary Committee, March 13, 1963, published in The Failing Newspaper Act, Hearings before the Subcommittee on Antitrust and Monopoly of the Judiciary Committee, U.S. Senate (1967, 1968, 1969), part 5, p. 2328. (Hereafter cited as Failing Newspaper Hearings.)

20. Ibid.

21. 2 FCC 208, 240 (1936) (dissenting opinion of Commissioner Stewart).

22. Broadcasting (February 15, 1937), p. 11.

23. Sterling, op. cit., pp. 230-31.

24. Tri-State Broadcasting v. FCC, 68 App.D.C. 292, 96 F.2d 564 (1938).

25. Heckman, op. cit., p. 380.

26. Order No. 79 (March 20, 1941), 6 Fed. Reg. 1580 (1941); and Order No. 79-A (July 1, 1941), 6 Fed. Reg. 3302 (1941).

27. Daniel W. Toohey, "Newspaper Ownership of Broadcast Facilities," Federal Communications Bar Journal 20:1 (1966): 47.

28. Heckman, op. cit., p. 380.

29. FCC v. Stahlman, 126 F.2d 124 (D.C.Cir. 1942).

30. Newspaper Ownership of Radio Stations, Notice of Dismissal of Proceedings, 9 Fed. Reg. 702 (1944).

31. Heckman, op. cit., p. 386.

32. Commission on Freedom of the Press, A Free and Responsible Press (Chicago: University of Chicago Press, 1947), pp. 43-46; see also the study sponsored by the Commission and authored by its vice-chairman, Zechariah Chafee, Jr., Government and Mass Communications, Vol. II, (Chicago: University of Chicago Press, 1947), pp. 653-66.

33. Heckman, op. cit., p. 387.

34. Mansfield Journal Co. v. FCC, 180 F.2d 28 (D.C.Cir. 1950); Scripps-Howard Radio, Inc. v. FCC, 189 F.2d 677, 683 (D.C.Cir. 1951).

35. Heckman, op. cit., pp. 390-93.

36. Toohey, op. cit., p. 53.

37. McClatchy Broadcasting Co. v. FCC, 239 F.2d 15, 18 (D.C. Cir. 1956) cert. denied, 353 U.S. 918 (1957); Tampa Times Co. v. FCC, 230 F.2d 224, 227 (D.C.Cir 1956); U.S. v. RCA, 358 U.S. 334, 351-2 (1959).

38. H.R. 9486, 84th Congress, 2nd Session (January 11, 1960).

39. 1 FCC 2d 394 (1965).

40. Broadcasting (May 13, 1968), p. 46.

41. Walter V. Kerr, "Newspaper-Broadcasting Monopolies: The Problem of Combinations," Saturday Review (October 12, 1968), p. 82.

42. Ibid., pp. 82-83.

43. Notice of Proposed Rulemaking, Standard, FM and Television Broadcast Stations - Multiple Ownership, FCC Docket No. 18,110, 33 Fed. Reg. 5315 (1968).

44. Broadcasting (December 9, 1968), p. 28.

45. Broadcasting (January 6, 1969), p. 21.

46. 16 FCC 2d 1 (1969); see also Louis Jaffe, "WHDH: The FCC and Broadcasting License Renewals," Harvard Law Review 82:8 (1969): 1693.

47. See "Three Billion Dollars in Stations Down the Drain?" Broadcasting (February 3, 1969), p. 19.

48. Broadcasting (January 27, 1969), pp. 25-26.

49. Amend Communications Act of 1934, Hearings on S.2004 before the Communications Subcommittee of the Committee on Commerce, U.S. Senate (1969), parts 1 and 2; see especially pp. 398-412.

50. Mass Media and Violence: A Staff Report to the National Commission on the Causes and Prevention of Violence (Washington, D.C.: U.S. Government Printing Office, 1969), p. 157.

51. Broadcasting (January 19, 1970), p. 30.

52. Chronicle Broadcasting Co., 16 FCC 2d 882 (1969); and Midwest Radio-Television, Inc., 16 FCC 2d 943 (1969).

53. Broadcasting (December 15, 1969), p. 9.

54. Further Notice of Proposed Rulemaking in Docket No. 18,110, 22 FCC 2d 339 (1970).

55. M. H. Seiden and Associates, Inc., Mass Communications in the United States - 1970, Summary Volume (Washington, D.C.: Research Conducted on Behalf of the National Association of Broadcasters, 1971).

56. Broadcasting (February 1, 1971), p. 34.

57. "Petition for Joint Study by Atlass Communications, et al.," filed June 11, 1970 in FCC Docket No. 18,110; see Broadcasting (June 15, 1970), p. 25.

58. 28 FCC 2d 662 (1971), issued shortly after the Seiden report was filed; see especially para. 10.

59. Minow's tenure is reviewed in Erik Barnouw, The Image Empire (New York: Oxford University Press, 1970), pp. 196-201.

60. Failing Newspaper Hearings, part 5, p. 2336.

61. Ibid., p. 2337.

62. Ibid., part 3, p. 1420; and Hearings on S.2004 (1969), part 1, p. 55.

63. Hearings on S.2004 (1969), part 1, p. 55.

64. Ibid.

65. Ibid., pp. 59-60.

66. George H. Litwin and William H. Wroth, The Effects of Common Ownership on Media Content and Influence (Prepared for the National Association of Broadcasters, 1969).

67. Stephen R. Barnett, "Cable Television and Media Concentration, Part I: Control of Cable Systems by Local Broadcasters," Stanford Law Review 22:2 (1970): 265.

68. 28 FCC 2d 662 (1971), para. 21.

69. Litwin and Wroth, op. cit., p. i.

70. Ibid., p. I-2.

71. Barnett, op. cit., p. 263.

72. 28 FCC 2d 662 (1971), para. 26.

73. Ibid., paras. 21-26; and First Report and Order in Docket 18,110, 22 FCC 2d 306 (1970), paras. 36-40.

74. Litwin and Wroth, op. cit., pp. I-1, II-8.

75. 22 FCC 2d 306 (1970), para. 37.

76. Harvey J. Levin, "The Policy on Joint Ownership of Newspapers and Television Stations: Some Assumptions, Objectives, and Effects," Statement in Docket No. 18,110 before the FCC (New York: Center for Policy Research, 1971), pp. 4-5.

77. "Petition for Joint Study. . .," pp. 3-4, quoted in ibid.

78. See NAB "Petition for Extension of Time to File Comments as to Further Notice, in Docket No. 18,110" (June 17, 1970), p. 3; ANPA "Motion to Extend Time for Preparation and Submission of Comments in Docket No. 18,110" (June 18, 1970), pp. 2-3; "Reply Comments of CBS in Docket No. 18,110" (February 27, 1969), pp. 3-4.

79. Levin, "The Policy on Joint Ownership. . .," p. 6; see also Harvey J. Levin, "Program Duplication, Diversity, and Effective Viewer Choices: Some Empirical Findings," American Economic Review 61:2 (1971): 81-88.

80. See the CBS-financed study by Gary A. Steiner, The People Look at Television (New York: Alfred A. Knopf, 1963), chaps. 5-6; and the NAB-financed report to the President's Task Force on Communications Policy by Herman W. Land Associates, Inc., Television and the Wired City (Washington, D.C.: NAB, 1968), chap. 2.

81. Levin, op. cit., "The Policy on Joint Ownership. . .," p. 6.

82. Ibid., pp. 52-54.

83. Failing Newspaper Hearings, part 5, p. 2349.

84. Broadcasting (February 3, 1975), pp. 23-24.

85. Quoted in Eugene Konecky, The American Communications Conspiracy (The People's Radio Foundation, 1948), p. 59.

In 1970, the same year that the Federal Communications Commission found itself surfeited with pleadings and studies defending and accosting its pluralist principles, the Westinghouse Broadcasting Corporation (WBC) was vigorously lobbying for a novel proposal. It strongly encouraged the Commission to take away from the networks a half hour each evening during prime time, and return it to its rightful owners, the local stations.[1] Without the networks to do their programming for them, the argument went, stations might be forced at last to begin servicing local expression in their communities, through more public-affairs and news programs.

The WBC proposal struck the Commission as a more direct route to its goal, compared with its circuitous ownership policies, which had grown more confusing and complicated in the latest cross-ownership rule-making. And the proposal could itself be construed as propluralist, in the sense that it would undermine slightly the control the network monopolies exercised over programming. Thus, over vehement protests from CBS, NBC and most network affiliated stations, the FCC adopted the prime time access rule, as it was called.[2] It became effective in the fall of 1971.

To the Commission's surprise, but not to WBC's, "the widespread early speculation that local news and public affairs broadcasts would be expanded to fill big chunks of the half-hours vacated by the networks was considerably short of the mark."[3] Instead surveys showed that most stations were filling the time slots with entertainment offerings, situation comedies ranking as the favorite choice.[4]

Many of these, in fact, came from a package of half-hour entertainment programs Westinghouse had prepared for national syndication just in time for the effective date of the prime time access rule. WBC's role in the affair did not pass unnoticed, and one envious program syndicator deplored the company's cynical opportunism. The

71

telefilm syndicator in this case was Screen Gems, which lacked an accompaniment of stations comparable to the chain of top market television properties owned by Westinghouse. The WBC stations provided that company with a guaranteed outlet for its syndicated entertainment package, and in the process, WBC stations were looking more and more like the launching pad for another network. The chief effect of the prime time access rule may have been, as Screen Gems claimed, that the ruling "has substituted one privileged monolith for another."[5]

The widespread unwillingness of station owners to comply with the spirit of the FCC's rule was predictably consistent with their record of unwillingness to provide forum opportunities and to service local expression in the information flow. That record was examined in the first part of our investigation.

And with some sophistication, the FCC seems to have anticipated that consequence. For the prime time access rule proved to be just the first of its efforts to move regulatory policy in the direction of requiring broadcasting's management to furnish programs it would not voluntarily provide. Before that rule had taken effect, the Commission had proposed another more "drastic" one, in the words of the industry's trade press.

Conditions favoring this new rule had taken shape since the FCC's denial of a license renewal to Boston TV station WHDH in 1969. That was the only successful license challenge before the Commission on record. Yet it had encouraged literally hundreds of other audience organizations to form in the years following, in order to challenge other owners. During this time the FCC had not awarded any challenger with a license stripped from an incumbent owner. Nonetheless, the industry felt aggrieved by the administrative and financial burdens imposed by license renewal hearings with adversaries present. Renewals had still followed, but not as automatically as had once been the case.

Accordingly, the Commission's newly proposed "drastic" rule offered stations an escape from the contentious aspect of the renewal ceremony. It provided that stations which evidenced "substantial service" would be assured the favor and support of the FCC at renewal time.

The Commission tentatively proposed to define "substantial service" for top 50 market VHF affiliates as follows: stations had to allocate 15 percent of air time to programs they produced themselves; and stations had to allocate 10 percent of broadcast time to news, and 5 percent to public affairs. If they chose, stations could satisfy both the local origination and informational requirements simultaneously, with the same programs. The Commission did not believe such expectations to be extraordinary, although it was aware that not all stations currently met the standards. For VHF-unaffiliated, or smaller market stations, however, the substantial service standards would be considerably reduced, or waived altogether. The FCC estimated that for the group of stations which would face the 15 percent local, 15 percent informational standard, a majority already topped the local requirement. In news allocations, however, this was true for only a third of the stations. And in public affairs, the Commission acknowledged that most stations would have to change their practices significantly: it was estimated that only 10 percent currently met that standard.[6]

Changing current practices in this area, of course, was exactly what the FCC hoped to achieve. In exchange for license security and a return to automatic renewal proceedings, the Commission hoped at the very most to encourage the industry to facilitate "community feedback" in its use of the airwaves; or alternatively, at least to compel broadcasters to air more public affairs and informational programs than they were willing to when the choice was exclusively theirs.[7]

The reaction to the substantial service proposal from audience organizations was that the guidelines were insubstantial. The Commission was criticized for failing to provide guarantees of community involvement in broadcasting, and for actually setting low, easily attainable performance standards. Increased percentages in all categories were recommended.[8]

The industry response could not have been more opposite. Representatives of the Storer stations, a major group broadcaster, reexamined and rearranged the Commission's own programming figures and announced their own conclusions: that on the basis of past programming, only 1 percent of the affected stations presently satisfied the proposed definition of substantial service. In a brief before the FCC, the Storer group ridiculed the idea that the Commission was offering the industry license stability. It said that interpretation "is defeated by a statement which says, in effect, that only five stations in the entire country merit renewal."[9] Doubtless there were some who wondered if even that figure might not be high.

In the wake of similar industry protests, the FCC seems to have backed off from its proposals. Several years have passed since the substantial service idea was first set aloft, with no effort on the Commission's part to resolve the matter. Some encouragement came from Congress in 1974 when the House of Representatives approved a license renewal bill that would have prohibited the consideration of ownership factors at renewal proceedings while authorizing the FCC to condition renewal chiefly on the demonstration of substantial service.[10] However, the bill died in that session.

But the whole turn in regulatory policy that is contemplated by rules such as prime time access and substantial service, is of more than passing curiosity. There appears to be an historic shift developing in regulatory approaches in favor of program requirements, however limited it may be in the sense that only very broad categories are stipulated in program allocation schemes. Paradoxically, this approach is emerging from the context of a communications system characterized on the one hand by a regulatory agency which has historically proved unwilling to trust its own principles; and on the other hand by an assortment of business monopolies which have a demonstrated unwillingness to do much more than add to their principal and service their own trusts.

Frustrated by decades of failure to achieve the business pluralism it imagined would provide responsive trustees for public telecommunications resources, the Commission shows signs of abandoning its historic, if narrowly construed, focus on providing for ownership pluralism and averting monopoly. It would retreat from an emphasis on the ownership structure; resign itself more fully to accepting the legitimacy of the prevailing control apparatus, which is certainly not characterized by pluralism; and substitute policy calculated to stimulate an increase in the information flow from those parties presently in control of the information machines.

If such a shift is in fact developing, it in itself raises fundamental questions that go to the core of pluralist philosophy and the meaning of a communications democracy. Legal, administrative and economic arrangements have given the communications corporations the power to determine whose interests are important, whether and how they are to be conveyed, by whom and when, over television. More than that, these arrangements have certified that power as a right. If the incentives which condition the exercise of that power have not provided forums for the vigorous exercise of free expression by diverse and antagonistic publics, how will the shape of the information flow be molded toward that end within the same power context, with its incentives unchanged?

This observation returns us to a problem raised in the introduction to this inquiry—the matter of the capacity of the telecommunications organizations to themselves represent diverse and antagonistic interests, needs and priorities in the communications flow. Given that access is their organizational privilege, and that they have not used it to facilitate forums for each and every interest group to speak for itself, it becomes critical to assess their own capacity to represent such diverse interests and to faithfully reflect such antagonisms. This is especially salient if new policies would require them to allocate more time to the representation of social intelligence and public affairs.

The questions raised here are honest, complex, and not resolvable through simple theorems and reassuring formulas. The biography of business pluralism should attest to that. Rather, such questions demand an examination of the organization of the telecommunications system, its control structure, and the incentives which condition its exercise of power.

We will begin that examination in the chapters that follow. And if the analysis is adequate to the task, we should emerge with a better appreciation of what the corporate interest is in our communications system, how business ownership shapes the information flow, and whether a communications democracy is at all possible in the context of highly developed business control.

NOTES

1. Broadcasting (March 16, 1970), pp. 10, 42-43; and "Network Television Broadcasting," 23 FCC 2d (May 4, 1970), p. 382.

2. Ibid.

3. Broadcasting (September 6, 1971), p. 16 ff.

4. Ibid.; Broadcasting (August 30, 1971), p. 47.

5. Broadcasting (May 31, 1971), p. 46.

6. Broadcasting (February 22, 1971), pp. 28-31; Broadcasting (March 1, 1971), pp. 31-33.

7. Ibid.

8. Broadcasting (December 20, 1971), p. 32; see also the revisions of "substantial service" proposed by the Office of Economic Opportunity in Broadcasting (September 13, 1971), pp. 28-29.

9. Ibid.

10. Broadcasting (February 25, 1974), pp. 7, 10; Broadcasting (March 4, 1974), pp. 32-33.

International alliances among great powers historically are marked with much fanfare. So it is with affairs of state. But the announcement in December, 1965, that a merger pact had been successfully negotiated between the International Telephone and Telegraph Company and the American Broadcasting Companies, Inc., was unmarked by the customary stentorian trappings. The principal signatories to the accord staged no joint press conferences and distributed no souvenir pens used to formally etch the agreement. Harold S. Geneen, ITT's president, was unceremoniously off in Europe surveying his bureaucratic divisions there. Leonard Goldenson, ABC's chief executive officer, remained at his company's world headquarters in New York, reflecting a business-as-usual air. So it is with the affairs of private power.

What the dignitaries, Geneen and Goldenson, had just forged was a private international alliance spanning five continents and joining the communications systems of scores of independent societies under their international corporate authority. With less visibility to the public eye was the union's inherent promise to strengthen the already well-established global ties among American military, industrial and communications interests.

Later, following swift FCC approval of this merger, Commissioner Nicholas Johnson wrote in dissent that "this particular transfer of broadcasting properties happens to be the largest in the history of the world."[1] But broadcasting is only a part, albeit a critical component, in the larger portfolio of international business activities that would be combined in an ITT-ABC merger.

To consider the implications of this merger, and the meaning of ABC's diversification and expansion into fields beyond domestic commercial broadcasting, one must return to ABC's beginnings as a corporation. But the saga of ABC is not exceptional among broad-

casting corporations, although it has its distinctive features. As a
multinational communications giant, it provides a prototypic example
of the imperatives of diversification and expansion which increasingly
govern the corporate use of our communications resources.

PHASE ONE: DIVERSIFICATION PRODUCES TELEFILMS

Even before it was a fully competitive television network, ABC
owners saw the wisdom in diversification. In 1951, at a time when
ABC still lacked regularly scheduled daytime television network programming, chief owner Edward J. Noble and president Robert Kintner
announced a merger agreement with Paramount Theatres, Inc. The
move established the first formal ownership linkage between broadcasting and the motion picture industry, and foreshadowed a major
turn in TV programming production and distribution.

Since the inception of TV, the movie industry had uniformly opposed any economic alliances with broadcasting. Television early
established itself as a live entertainment medium available free with
the purchase of a set, and the apparent vulnerability of the filmed entertainment industry seemed thereby to require a protective posture.
Paradoxically, the only previous television penetrations of Hollywood
had been NBC and CBS contracts for newsfilm, and these had been
short-lived.

The FCC finally approved the ABC-Paramount transaction in
mid-1953, and Paramount executive Leonard Goldenson assumed the
presidency of the newly merged corporation. With a beachhead in the
film world thus established, the Goldenson entree paid off rapidly,
producing within a year a signed contract with Disney Studios for the
first Hollywood productions for television. The news was quickly
followed by announcement of a second far-reaching agreement: ABC
had contracted with Warner Brothers for a package of forty action-
adventure telefilms for the 1955-56 season.[2]

Midway into 1955 the signs of economic change were already
becoming clear. ABC-TV, under the Goldenson-Kintner leadership,
reported gross billings had leaped 68 percent above the previous
year's total. The reason—Disney Studios.[3] By 1956, with the Warner
Brothers package unleashed, the shift in production from live entertainment to telefilms was on. Media historian Erik Barnouw tells us
that "countless stations reduced staffs, closed expensive studios, and
took up round the clock film projection."[4] In other words the local
programming staff gave way to the studio projectionist, and local programming was abandoned for network fare, Hollywood telefilms.

Paralleling the shift in production from New York studios and local stations everywhere to the central Hollywood complex was a simultaneous diffusion of management. NBC and CBS followed ABC into telefilms and turned for guidance to the ABC leadership. Kintner moved over to the NBC presidency in late 1956 and James Aubrey, another ABC pilot, moved first to the vice-presidency and then became head of the CBS network. The year 1959 found ABCBSNBC programming an endless diet of telefilms under the supervision of the original ABC management team. Thus was the initial ABC model disseminated.

The telefilm phenomenon, pioneered in ABC's first diversification drive, was thereafter to function as the chosen instrument of an alliance between the media and consumer goods industries. The telefilm allegedly provided an ideal package for the commercial message. This was presumed to be especially true with respect to the emotional environment the telefilm defined for the viewer: it evoked and reinforced the same structure of emotional needs which produced an ego and emotive involvement with consumer goods. Or so the social psychological consultants informed the broadcasting and advertising industries. This symbiotic relationship between telefilms and commercials was widely touted by the trade press and advertising psychology journals from the beginning of the telefilm ascendancy.[5]

Regardless of how compatible or complementary the two were in fact, nevertheless the telefilm became the vehicle for a massive infusion of consumer goods advertising into television, and ultimately the living room. TV advertising mushroomed. From 1954, the last year before the introduction of telefilms, to 1957, when conversion was complete, the number of television stations operating grew by 32 percent, while advertising time sales more than doubled that pace, soaring 65 percent. Telefilms had ushered in an advertising gold rush.

All of this helped to draw the network broadcasters and the national consumer goods manufacturers into a closer alliance. From the beginning, broadcasters had courted advertisers to finance their operations, but not simply any advertisers. Rather, the broadcasters were prepared to limit paid access to commercial interests. The latter discovered that with telefilm entertainment the medium was an important resource in stimulating their own growth and development. Accordingly, the consumer-goods manufacturers proved to be equally prepared to transfer to the owners of this technology much of their own wealth in exchange for this exclusive access. Without the telefilm or its equivalent, the development of television communications as a consumer delivery enterprise for retail manufacturers might not have been achieved as speedily or as absolutely.

However, the expansion of the consumer-goods/television alliance beyond American borders required a more complex strategy, even given the advantages inherent in the telefilm. And these were not negligible. To create programming for domestic and international markets simultaneously, a formula was developed: emphasize filmed action in defining situations and conveying meaning as much as possible, reducing in turn any dependence upon dialogue. The formula virtually guaranteed that the product would be interchangeable among cultures, or markets, with only minimal dubbing procedures. It also meant that telefilms would be case overwhelmingly in an action-adventure-crime-mystery format, with all of the universalistic appeal inherent in any simplistic, Manichaen storyline.

PHASE TWO: MULTINATIONAL EXPANSION

The first evidence of ABC's intentions to concentrate on foreign empire building came in 1959 with the establishment of ABC International. Since the advent of telefilms, the networks and Hollywood film companies had moved into the export market with the bulk of sales going to Canada and Great Britain.6 But with the ambitious title of "Worldvision" for its operations, ABC, as became clear in developments over succeeding years, planned to put together not simply an international sales operation, rather, it proceeded to create for itself a global system of ownership, management, production, distribution and advertising services—in short, a concentration of broadcasting control internationally which surpassed the then level of industry integration domestically.

The pieces in ABC's puzzle quickly fell into place. In the fall of 1959 the network quiz scandals erupted with Charles Van Doren's confession that his money-winning performance had been rigged. As one after another former contestant came forward to repeat the story, ABC, joined by the other networks, quickly moved to use the incident to justify further network control over program production.

Quiz shows and most other program fare had been produced under contract with sponsors, not networks. The sponsors then purchased network program time which they were free to fill with their own shows. The scandals thus proffered a cover for the networks to reverse the lines of authority and the flow of profits. In a magnanimous show of public responsibility, the network executives promised to take firm control of their program fare by contracting for production themselves, and selling spot time to sponsors.7 This assured them new profit participation in shows aired domestically and later exported for foreign syndication.

Several months later the policy was extended to provide for a network monopoly on production of informational programming. The opportunity for this was precipitated when, in early 1960, each of the three networks refused to air a sponsor-produced documentary. ABC, first to justify its stand, explained: "The standards of production and presentation which apply to a professional network news department would not necessarily apply to, for instance, an independent Hollywood producer."8 The other networks followed suit, but the ABC position was the most transparently about-faced and embarrassing, with the opposition to Hollywood production being led by the network most closely linked to Hollywood. This seeming contradiction had its resolution, to be sure, in the control it afforded ABC in establishing the focus and direction of all informational programming it might broadcast. And in the process ABC had swiftly accomplished the concentration of control of production, distribution and profit participation for all programming, informational and entertaining, it aired at home and was eager to air abroad.

In the wake of this latest move, Goldenson contracted with one of the tainted outside production companies for a series of documentaries, a step which embarrassed John Charles Daley (his vice-president for news) into resigning. The locus of the series, curiously, was Latin America, particularly the countries of Venezuela, Costa Rica and Cuba.

From ABC's viewpoint these three countries represented, respectively, the corporate dream realized, its next frontier, and the dream-turned-nightmare. For while ABC was occupied with domestic maneuvers to concentrate program production, distribution and profits in its own hands, it was also busily engaged in plans to replicate its scheme in foreign areas, particularly Latin America. And although ABC's Worldvision was just celebrating its first birthday in 1960, the division's international successes were already evident.

In Venezuela ABC was part-owner of three of the country's 14 television stations, and managed programming on another eight through its Worldvision contractual arrangements. ABC thus dominated programming on 11 of the country's 14 television channels.

Costa Rica, on the other hand, had no television stations at the moment. But as one of six Central American countries, it was a target of ABC's goals for the future of that area.

ABC's focus on Central America is instructive. Prompted by the fear of socialist revolutions following the example of Fidel Castro's lead in Cuba, U.S. government officials and businessmen began in 1960 to press hard for the formation of an economic community among the Central American nations. The economic motives underpinning American business interest in this independent community became clear when the agreement signed in Costa Rica that same year provided for "U.S. businesses to be established in one Central American country and to sell their goods in all the countries, without paying any tariffs."9 But the rapid development of foreign markets for American consumer goods, even those manufactured by U.S. sub-sidiaries abroad, was dependent upon another factor—the simultaneous development of commercial television in these countries. Otherwise U.S. corporations would lack the accustomed cummunications appara-tus necessary to sell their products. In other words, the industrial barricade against Cuban socialism needed the cement that only com-mercial broadcasting safely tended by American hands could provide.

And so it happened that also in 1960, ABC International of New York advanced funds for the establishment of television stations in each of the five countries which were then forming the Central Amer-ican Free Trade Area. With this investment foothold, ABC subse-quently announced the creation of the Central American Television Network (CATVN).[10] The fact that this network was organized with-out any arrangements or even plans for live program exchanges among the member stations made clear the role ABC had created for itself. Essentially the network was a public relations label for the international program sales and commercial advertising opera-tions of ABC International. The ABC-owned stations in these coun-tries would be linked together in the CATVN through the facilities of ABC's Worldvision, which would "sell" programs to the stations and handle their sale of commercial advertising.

The provocation which had served to assemble this next frontier and which, at the same time, threatened its development, was the eco-nomic and political change occurring in Cuba. This third site of ABC's documentary filming, Cuba was the corporation's dream-turned-sour, a commercial communications nightmare.

The Cuban medium had developed earlier than in other Latin countries, and Cuba was the first country in the world to make tele-vision available to the entire population.[11] Broadcasting was thor-oughly commercialized and functioned with only nominal government supervision. As one enthusiast of U.S. broadcasting's overseas ad-ventures has written,

> In pre-Castro days, Cuban TV was a highly commercialized operation, controlled by businessmen who were aware of the political value of stressing light entertainment programs and of shying away from anything that might irritate the Batista regime. The medium was a bland melding of local talent shows and American serials.[12]

In other words, Cuban television had developed along the lines of the American model and, predictably, had proved a lucrative market for American telefilm exports. The writer goes on to note "At the time of the Fidelista takeover, there were sixty-two American serials on Cuban TV, heavily weighted toward comedies and cowboys."[13] In

short, the only prominent difference between television in Miami and
Havana was the language spoken; the two looked the same.

In 1960 the change in governments produced a different use of
the media. Fidel Castro and the United Party of Socialist Revolution
nationalized the television stations and initiated their own programming. They also closed the door to American corporate commercial
sponsors and halted the import of American telefilms. It seemed that
the commercial model of broadcasting conceived by American business entrepreneurs and prepared for export around the world was
suddenly imperiled, should the Cuban example have found a following
in other Latin capitals.

This perhaps was the nature of the Latin American turmoil
which prompted Goldenson's decision in 1960 to produce a documentary series on the area, as he dispatched camera crews to Venezuela,
Costa Rica and Cuba. But between the camera's focus on a militant
Castro and the aggressive American diplomatic efforts in Costa Rica
to erect an industrial barricade, there was no mention of ABC's con-
cordance in the struggle.

Once its investments in the Central American Free Trade Area
(Costa Rica, El Salvador, Guatamala, Honduras, Nicaragua) proved
successful, ABC pushed ahead with further station construction and
acquisition to add to facilities it already owned in Canada, Argentina,
Venezuela and Panama. By the mid-1960s stations owned by ABC were
operating from Chile, Columbia, Ecuador, Honduras, Bermuda, Lebanon, Japan, Ryukus, the Phillipines and Australia, bringing the total to
17 countries where ABC had ownership interests in television.* And
to the CATVN, ABC had added the Arab Middle East Network, also
under the aegis of Worldvision.[14] So rapid and unchallenged was the
rise of ABC International and Worldvision that the corporation was
able to claim in its promotional copy that "no other single television
source exists that has anything remotely resembling this global broadcast service."[15]

Worldvision operations were brought to bear on commercial
television systems wherever direct financial investments in stations
by ABC were not feasible. An agreement with Worldvision, however,
provided ABC with many of the same control mechanisms that come
with actual ownership. The Worldvision contract transferred to ABC
responsibilities for buying programs, selling sponsorships, and scheduling broadcasts. The Worldvision service " provides that each

---

*In 1968 ABC International added the Latin American Television
International Network Organization (LATINO), modeled after CATVN
and including the countries of Venezuela, Ecuador, Uruguay, Argentina,
Chile and Mexico.

Worldvision station must make Class A time available within eight
weeks for any program presold by ABC. For example, ABC can sell
Batman to an advertiser and then place Batman along with designated
commercials in any Worldvision country where the advertiser wants
it to appear."[16] By 1966 Worldvision reached 20 million TV homes,
an impressive 60 percent of world TV homes in areas with commercial
television outside of the United States.[17] By 1969 ABC was distribut-
ing almost 900 programs to more than 90 countries.[18]

TELEVISION: THE AMBASSADOR OF AMERICAN MILITARY-INDUSTRIAL INTERESTS

Out of ABC's expansion policies, pioneered in Latin America
and eventually generalized to other continents, a recurrent pattern
emerges.

The key to the pattern is provided by economics, or more specifically, profit. In American broadcasting, profit derives from paid
access by consumer goods manufacturers, especially those which dis-
tribute over large geographical areas and thus appreciate the benefits
of mass advertising to stimulate demand. In foreign countries the development of broadcasting has often approximated this same model,
even to the extent of a dependence upon American manufacturers. As
the example of ABC's move into Central America suggests, entree to
foreign soil by American broadcasters is fundamentally conditioned
upon expectations that other American businesses will also enter.
This is so because U.S.-based consumer goods and service industries
are inclined to rely heavily on mass media advertising of the kind that
television uniquely provides. In effect, the future potential or existing
availability of American business operations abroad provides a guarantee of continuous financing to American broadcasters eager themselves to expand abroad. The result is that American industry and
commercial broadcasting are typically exported in tandem, as we saw
was the case in the Central American region.

Another important element in the pattern not considered thus
far is the role adopted by diverse sectors of the U.S. government,
which one could only characterize as supportive of American broadcast (and business) hegemony in overseas countries. The core of the
government's interest in the matter arises from its own very substantial broadcast operations already entrenched on foreign soil,
somewhat precariously because of their obvious military and political
propaganda significance. The Department of Defense, for example,
has operated for some time a broadcasting network encircling the
globe with 38 television and more than 200 radio transmitters in place

by the early 1960s.[19] Established ostensibly to provide Americans
abroad, particularly military forces, with information and entertainment from home, the stations also reach a foreign audience. The State
Department reports with pride that this native audience outnumbers
the American overseas audience by a 20-to-one ratio.[20]

The civilian arm of the government charged with broadcasting
propaganda abroad, the U.S. Information Agency (USIA), depends at
least as much as the American military upon cooperative relations
with foreign governments, and broadcasters particularly. For although
the USIA operates 43 powerful domestically based radio transmitters
and an additional 59 radio transmitters located in overseas countries,
its television program material is aired over foreign stations (both
private- and state-owned).[21] Of course when foreign stations are subject to American business influence either through direct investment,
Worldvision contracts, or other congenial arrangements, USIA access
is facilitated considerably. Predictably, the programs meet with little
resistance. Since 1965 they have been aired annually by more than
2,000 television stations.[22] As of 1972, the USIA audience in 97 countries was estimated to be more than 900 million a year.[23]

The strategic importance of this government communications
network abroad, including both its civilian and military wings, to
American foreign policy has been clearly explained in countless federal reports and by a succession of government spokesmen. In 1964
the House Foreign Affairs Committee advised as follows:

> Certain foreign policy objectives can be pursued by dealing
> directly with the people of foreign countries, rather than
> with their governments. Through the use of modern instru-
> ments and techniques of communications it is possible to-
> day to reach large or influential segments of national popu-
> lations—to inform them, to influence their attitudes, and
> at times perhaps even to motivate them to a particular
> course of action. These groups, in turn, are capable of
> exerting noticeable, even decisive, pressures, on their
> governments.[24]

In 1971 USIA director Frank Shakespeare* voiced continuity for this
policy with the following assessment of the USIA role:

---

*Prior to his appointment as USIA director in 1969, Shakespeare
was in charge of CBS's foreign TV investments, and the worldwide
distribution of CBS programs to foreign TV stations. During 1968 he
was a principal advisor for the media compaign of Richard Nixon. He
resigned his USIA post in 1973 to become the executive vice-president
of the Westinghouse Electric Corp., and second-in-command of WBC.

> I am hopeful that with the massive development of com-
> munications we shall be able to live in a world where
> free access to the minds of people everywhere will en-
> able us to achieve our goals through competition in
> thoughts, rather than in armaments and power pres-
> sures.[25]

The special circumstances which permit the U.S. government's "free access to the minds of people everywhere" arise in part from the presence abroad of American industrial and broadcasting corpo- rations. These circumstances are reproduced daily around the globe through a tidy structure of mutual support and self-interest. They be- gin, for example, with the preparation of USIA versions of the news, under the jurisdiction of USIA television officers assigned to each American embassy. The television officers' arrangements for com- mercial sponsorship of the news program are often with an American- based multinational corporation. The corporation in turn contracts for local broadcast in prime time over the facilities of an American- owned or -operated local station. Thus the mutual penetration by American enterprise and broadcasting facilitates the U.S. govern- ment's communications objectives through a structure which neatly diminishes most viable means of native jurisdiction or intervention. Consequently, USIA news programs have been a regular feature on more than 100 television stations in Latin America alone.[26] And the efficiency of the structure in circumventing native control produces other dividends—most USIA programs are never identified as such in most of the countries in which they are shown.[27]

Perhaps the most salient payoff for the U.S. military and propa- ganda arms lies in the neutralization of what would otherwise be an antagonistic force, the native broadcasting industry. For the fact of American control, in the form either of ownership or programming authority or both, not only provides for easy U.S. governmental ac- cess as we have seen; it also dilutes and resolves otherwise natural contradictions between domestic and foreign use of telecommunica- tions resources. Where a native, nationalistically inclined industry would demand protection from foreign government use of the domestic spectrum, we find instead that the native communication system is itself heavily supplied and managed by foreign interests. Consequently American station owners, program suppliers and directors, aware of the degree to which the American government has installed them and protects their investments, prove themselves a reliable ally to the ad- vance of Defense Department and USIA transmitters and broadcast programs in nations overseas.

The U.S. government and its broadcasters, to be sure, have long understood the advantage of securing invisible access abroad for

business and military propaganda purposes—along with the other advantages which accompany American penetration of foreign broadcasting institutions. That is why the Communications Act of 1934 ex-
pressly prohibited any foreign ownership interests in American
broadcasting operations.

On the other side of this exchange relationship, there have been
a number of government programs which have functioned over the
years to open, develop and maintain foreign television markets for the
benefit of American broadcasters. The range of programs includes
extending incentives to overseas countries, financing construction of
broadcast systems, and insuring American investments abroad. Most
such programs are billed for public consumption as foreign aid.

The earliest government-sponsored programs were begun in the
1950s by the State Department, just at the time when telefilms were
proving to be easily exportable. Under the lofty title of "cultural exchange programs," federal monies were made available to finance
visits by foreign officials and executives to fraternize with U.S. broadcasters and observe the workings of American television. The exchange dimension of this program amounted to taxpayer-financed
visits abroad by American experts, enabling them to size up future
markets first-hand.[28]

When incentives like this have failed to produce quick invitations
to American commercial broadcasters, the U.S. government has occasionally offered to absorb the costs itself of building foreign television systems. In 1959 for example, the United States agreed to finance construction of the entire United Arab Republic (Egypt and
Syria) TV system, which began telecasting the following year. With
the financing of additional stations, Jordanian audiences were brought
into the project. Following the fulfillment of U.S. government commitments, ABC International announced the formation of the Arab
Middle East Network. It consisted of TV stations serving Syria and
Jordan as well as Kuwait (largely the work of American engineers),
Iraq (constructed in part by American technicians), and Lebanon
(where the Ford Foundation, Time-Life and ABC had each been active
building stations).[29] The network functions as a one-way conduit for
American programs to the Middle East, rather than as an intercon-
nection for program exchanges among Arab stations. The latter,
according to one account, has long been a goal of Middle East leaders;
yet ABC has been able to exact more cooperation for its network than
have the Arabs for theirs.[30]

Similarly, in 1965 the U.S. Corps of Engineers was assigned the
task of opening up new television markets in Saudi Arabia, building a
national system there. Once under way, however, the project was
handed over to NBC International for further development and expansion.[31]

Another mode of U.S. subsidy to commercial broadcasting expansion has been conducted under the banner of "educational TV assistance," through programs begun in 1961. Since that time the Agency for International Development (AID) has provided free TV receivers to the school systems of some third world nations, and launched other projects aimed at introducing television to the foreign classroom.32 Developing nations have proved eager to partake of this electronic age technology, not alone for its symbolic value, but to realize the promises told by American advisors and detailed in UNESCO studies, of instant revolutions in education via television. Such TV set giveaways and allied projects, while cheaper than financing station construction, have the advantage of introducing the medium to millions in the youngest age groups for the first time. AID education projects reduce the initially overwhelming problem of set penetration that exists in countries with populations too poor to purchase sets on their own. And they establish viewing patterns and an audience for TV in the very age groups which offer the highest potential for future set purchases and consumer expenditures. Thus to entrenched or watchful American broadcast investors educational TV assistance may be notable chiefly for its function in speeding the introduction and proliferation of commercial broadcasting.

Communications satellites, developed with more than $20 billion in public funds, offer yet another example of federal subsidies facilitating the designs of American-based multinational broadcasters.33 Beginning with the administration of President John Kennedy, internationally dispersed ground stations necessary for satellite transmission have been built by U.S. communications corporations with government funds. Satellites provide American commercial broadcasters with the technology for the first time to reach all of their international audiences simultaneously. The international space satellite network has been launched by the Communications Satellite Corporation (COMSAT) and its international affiliate INTELSAT, with control of both organizations shared by the U.S. government and American corporations. The satellite system has been developed in time for the instantaneous live transmission of American spectaculars like the moon landing to Worldvision and other stations included in the global American network.

U.S. government each lubricating the growth and facilitating the goals of the others' foreign expansion, a structure of mutual interest and dependency emerges. As this troika faces outward beyond American borders its separate interests begin to converge. Multinational expansion of American industry is conditioned upon the dissemination of the American commercial broadcasting system. U.S. government agencies promote diffusion of the commercial model, provide some aid for the installation of American broadcasters, and guard their

investments. Entrenched abroad, this communications system serves
as the ambassador of the American business system and the passport
for government propaganda. The dynamics of the system evidence an
alliance rooted in interdependency and manifested in reciprocal support among American communications, industrial and political institutions.

While this kind of international expansion has been a special
concern of ABC, the ambassador role of American telecommunications
abroad is a product of expansion drives by numerous American broadcasters, including particularly CBS and NBC, each with a distinctive
and complementary approach. It is also a product of expansion interests of American industrial corporations and U.S. government agencies, as we have seen.

But this seeming digression from ABC's corporate biography
is undertaken above not to make an analysis of modern day colonial-
ism or imperialism, although the evidence in that regard is provoca-
tive. It is rather to direct attention to two more critical points about
the effects of business ownership on the use of telecommunications
resources. The case is often that what seem to be normal, even natural processes assume a different, and therein revealing, appear-
ance in unfamiliar settings. So it is with the multinational expansion
of the broadcasting business. With the extension of American broadcasting beyond American borders, we see the outline of its structure
and dynamics clarified. For the business use of communications resources abroad is rooted in the business use of that resource at home.

What is seen first of all here is a mutual interest between broadcasting organizations and the business sector generally. Their common purpose, which in the case of foreign expansion requires considerable coordination, is to use telecommunications technology and
shape the communications flow to better procure customers; for industrial products and services and; in the present case, for foreign
imperial interests as well. Communications here involves no exchange, but a one-way flow, which originates in New York or Hollywood
and is exported to foreign cultures in much the same fashion as it is
exported to local American stations. And the produce itself is not so
much communications as it is audiences. Business ownership of communications resources, in alliance with other business interests,
functions to fashion these resources chiefly into consumer delivery
systems.

The second point of significance is that in the course of expansion, we observe the intensification of these institutional interdependencies between broadcasting and the larger business sector. Multinational expansion of manufacturing and retailing businesses virtually
requires that broadcasting's owners become their willing partners
and comrades. Expansion invites broadcasting's owners to become

attuned to common interests with other businesses, and the possibilities for reciprocal and mutually beneficial actions.

In the long run, this prepares the way for another phenomenon—the blurring of boundaries between the control structures of broadcasting and those of the corporate industrial system. One may reasonably assume that the more these institutions come under common control, the more incentives exist to identify and exploit their shared interests, within the limits defined by their separate functions. Yet the predominant use of telecommunications resources for consumer delivery purposes makes it difficult to speak confidently of the separate functions of a business-owned broadcasting system. Moreover, multinational expansion seems to encourage this blurring of boundaries, as does the increasing diversification within the broadcasting industry itself.

Reflective of an emerging institutional synthesis is the carefully planned melding of two corporate superpowers, ABC and ITT. The preceding analysis began with the announcement of their merger in December of 1965.

PHASE THREE: SYNTHESIS THROUGH DIVERSIFICATION

The International Telephone and Telegraph Company is the world's ninth largest corporation in number of employees, spanning all continents, with holdings in over 50 countries. It is a multinational conglomerate centered in communications and electronics in both consumer and hardware lines, from manufacturing to merchandising to servicing. While its 275 factories range worldwide, its operations, like those of ABC, are concentrated most heavily in Latin America. ITT is the second largest stockholder in COMSAT, and one of only four common carriers with global authority to lease satellite circuits and land lines to ground stations.* In each of these areas lie the points of contact and grounds for integration between ABC, multinational broadcaster, and ITT, multinational industrial conglomerate.

From ABC's position, the most immediate benefit (if one can be singled out), is the internal access to satellite circuits and ground stations that the merger would afford. This would bring to fruition the second element in ABC's bilateral program for world communications dominance. Detailing this program in its 1961 Annual Report,

---

*The assessments of corporate stature included in this section are effective for the period covered by the merger proceedings, December, 1965 to January, 1968.

the company noted that complementary with its overseas station investments and Worldvision operations, ABC International was also "continuing its activities to establish the associations and partnerships which will be necessary to translate satellite transmissions from a scientific curiosity into an effective and profitable worldwide television system."[34] ITT ably satisfies these conditions.

Owing perhaps to its emphasis on foreign expansion, ABC has never developed in the areas of electronics research, engineering and manufacturing on a scale like that of its sister networks, NBC and CBS. Consequently it has been excluded from controlling positions as new telecommunications technologies have been developed, rendering previous systems obsolete. In contrast, NBC, owned in common with RCA, has been advantaged in satellite communications since RCA, like ITT, manufactures satellite equipment, is a stockholder in COMSAT, and is an authorized global lessor of satellite services. Thus NBC-RCA is in a unique position to profit from increased international networking, in turn threatening to cloister ABC, the first corporate power to assemble a comprehensive international network of stations. In counter moves, ABC early pressed for an emphasis on the development of satellite ground stations in Latin America, and challenged COMSAT's commercial satellite monopoly, petitioning the FCC for authority to launch and operate a satellite system of its own.[35] Failing on both counts, ABC turned to the shelter of ITT.

For ABC, other distinct advantages would flow from the merger as well. ITT's Geneen stated "ITT will help most in international areas . . ."[36] No doubt this would be particularly true as the merger would join major multinational consumer goods industries, mainstays of advertising support, with multinational broadcasting, a mainstay of successful consumer goods marketing. Thus the reciprocities and interdependencies previously noted between these institutions would be cemented, enhancing coordination by consolidating their operations under common corporate control. As the trade press noted, "ITT's substantial manufacturing and sales in foreign markets, particularly in Latin America, must work to the advantage of ABC's developing broadcast interests there."[37]

At home and abroad, the merger with ITT's diversified manufacturing and service operations would reduce further the differences in vested interests between industrial marketing and communications. Specifically the merger would provide ABC with new ownership interests in the manufacturing of television sets, record players, refrigerators and numerous other electric appliances, as well as in consumer finance companies, life insurance companies and automotive rentals. With investment in these operations comes a special interest in the trade regulations which govern the marketing of these goods and services in each of the countries where ITT does business.

Of particular import to ABC is the tie to the defense and aero-space industry included in the merger package. ITT, an annual member of the inner circle of top 30 Pentagon contractors, received 40 percent of its U.S. income from defense and space contracts in 1965.[38] Accordingly, ABC would be assured a privileged inside track in tele-communications and electronics research and development (all financed by public funds), erasing the corporation's historic dependent status in this area of engineering and technology. Consequently ABC would at last be in a position to gain promotional advantages through televised space launches like those enjoyed by NBC and CBS, longtime beneficiaries of NASA contracts.

On ITT's side of the merger equation the gains are equally impressive. ITT's most compelling concern is to reverse its traditional ratio of U.S.-foreign investments, an imbalance which undermines the American identity of the corporation, which ITT management believes imperils in turn U.S. government advocacy and guaranty of ITT's foreign interests. Such advocacy often plays a strategic role in ITT's international ambitions. From 1959 when Geneen assumed control of the company, ITT's U.S. income was boosted from 21 percent to 50 percent of the corporate total at the time of the merger announcement.[39] The acquisition of ABC would thus assure attainment of Geneen's proclaimed goal of a 60 percent U.S.-40 percent foreign-income ratio, while at the same time uniting ITT with a highly visible American corporation.

ITT's reliance upon U.S. diplomatic actions supportive of its interests is a matter of public record and an achievement Geneen has personally emphasized in his reports to stockholders. When, for example, ITT's Brazilian interests were nationalized in 1963 (but only after the company was able to exact considerably more favorable terms of compensation than were offered originally), Geneen explained to a gathering of stockholders "Recent events in Brazil have drama-tically supported an improving turn of events there, and I draw your attention to the significance of President Johnson's recognition of the new Brazilian government. . . ."[40] In the following year when ITT was locked in a struggle with the Japanese government over tax claims, the corporation was similarly successful in getting the State Department to intervene with the Japanese ministry on its behalf.[41]

As Geneen himself stresses, the security of ITT's operations is "considerably strengthened by better government-to-government understanding on the part of our own government and the governments of other countries, particularly in Latin America."[42] And it is par-ticularly in Latin America where ABC and ITT operations overlap, with ABC broadcasting from eight of the 12 countries in which ITT has investments. Adding ABC's domestic network (reaching over 93 percent of U.S. TV and radio homes) to these and other foreign holdings, gains access for ITT to audiences worldwide. Thus the merger secures this strategic weapon for ITT's arsenal for aggressive international maneuvers.

Reflecting these intentions was the conspicuous attendance at the merger proceedings of ITT's corporate foreign policy executive, John McCone. McCone was the only official present to testify and advise at the FCC's hearings in 1966, besides Geneen and Goldenson. Brought to ITT by Geneen, McCone has a much-coveted reputation in some circles for ingenuity in foreign maneuvers, reaching back to the days when he was head of the CIA, the position he left to join ITT. Geneen has noted proudly in this regard that ITT

> . . . has in its time met and surmounted every device employed by governments to encourage their own industries and hamper those of foreigners, including taxes, tariffs, quotas, currency restrictions, subsidies, barter assignments, guarantees, moratoriums, devaluations—yes, and nationalizations.[43]

At least one of the devices ITT favors to impose its objectives on governments was unveiled during FCC hearings on the merger. As the hearings proved more arduous than ITT had anticipated, officials (including the senior vice-president for public relations) endeavored to exploit their personal relationships with reporters. Associated Press, United Press International and New York Times correspondents covering the merger case were asked to write more sympathetic portrayals of ITT's interests in the merger. When this initial application of pressure failed to secure newspaper accounts of the proceedings more to the satisfaction of ITT, the executives grew desperate, resorting to threats in an attempt to force journalists to change their stories.[44]

Whether or not ABC could be expected to turn over its worldwide communications apparatus to the service of similar objectives in future times and places, can be guessed perhaps by Geneen's own appraisal of ABC's interests.

> I have faith in ABC's management. It is the youngest and hungriest network. I've looked at it and its a smooth operation. . . . I expect it to fit in easily with ITT.[45]

In the evolution of the broadcasting business, the ABC-ITT merger agreement is certainly one of the most ambitious corporate alliances. The pact aligns worldwide communications systems with worldwide consumer manufacturing and service industries under the sovereignty of a single U.S. corporate power, and within the shield of

American diplomatic and military interests abroad. Surprisingly, but predictably perhaps, little mention of these aspects occurred at the FCC's swift one day hearing on the merger. There ABC attempted once again to trade on its "weak sister, third network" image, with obvious success.* In 1966 the Commission approved the merger in a move Commissioner Johnson said "makes a mockery of the public responsibility of a regulatory commission that is perhaps unparalleled in the history of American administrative process."[46]

But the matter did not rest there. The Justice Department, provoked by the FCC's abrupt handling of the merger, appealed to the Commission for reconsideration. In a spirit of accommodation, ritualistic hearings were reopened in 1967, after which the FCC, for the second time, issued its stamp of approval. Only when the Justice Department proved insistent by announcing plans to take the merger to court, did the parties grow weary. Shaken by the public exposure of its efforts to tamper with press coverage of the merger, ITT announced it would not further test its victories in the federal courts. Thus in early 1968, more than two years after ABC and ITT had concluded their merger negotiations, ITT withdrew from its contract with ABC.

The international integration of massive communications systems and technologies with defense and industrial complexes was thus never culminated in the particular case of ABC and ITT. But what may have been temporarily stalled abroad, for ABC's interests at least, was by the 1970s already an advancing reality at home.

---

*For the purpose of winning support for its interests in hearings like these, historically, ABC has fostered an image of a network struggling to keep abreast of its earlier established competitors, NBC and CBS. The trick involves confining the image to regulatory arenas, and keeping it from spreading to advertising and financial circles where the prophecy could prove self-fulfilling. Thus in regulatory proceedings ABC is given to lamenting its tardy entry into the network race, its consequent failure to secure affiliate stations in rural, small town, two-station markets, and its being left with residual UHF stations in the mixed, middle-sized markets where its rivals earlier grabbed affiliations with the only available VHF licenses. Advertisers, however, hear of ABC's reach into more than 93 percent of American TV homes, ABC's frequent first-place audience ratings in the top 30 markets where the bulk of the advertising dollar is spent, and the network's pioneering successes in telefilms.

NOTES

1. ABC-ITT Merger, Memorandum Opinion and Order, Dissenting Opinion of Commissioner Nicholas Johnson, 9 P & F Radio Reg. 2d 46 (December 21, 1966).

2. Erik Barnouw, The Image Empire (New York: Oxford University Press, 1970), p. 62.

3. "ABC-TV Is Making Its Move for Power Bid in TV Race," Broadcasting (June 20, 1955), pp. 27-28.

4. Barnouw, op. cit., p. 65.

5. Ibid., pp. 67, 81-82.

6. Wilson P. Dizard, "American Television's Foreign Markets," Television Quarterly 3, no. 3 (Summer 1964): 61-62.

7. Barnouw, op. cit., p. 150.

8. Quoted in ibid., p. 159.

9. Jon Frappier, "U.S. Media Empire/Latin America," NACLA Newsletter 2, no. 9 (January 1969): 4.

10. Ibid.

11. Wilson P. Dizard, Television: A World View (Syracuse: Syracuse University Press, 1966), p. 52.

12. Ibid., p. 134.

13. Ibid.

14. Ibid., p. 98.

15. Quoted in Harry J. Skornia, "American Broadcasters Abroad," Quarterly Review of Economics and Business 4, no. 3 (Autumn 1964): 16.

16. Ralph Tyler, "Television Around the World," Television Magazine 23, no. 10 (October 1966): 33.

17. From an advertisement for ABC International-Worldvision, in Television Magazine 23, no. 10 (October 1966): 61.

18. Andres R. Horowitz, "The Global Bonanza of American TV," More 5, no. 5 (May 1975): 6.

19. Herbert I. Schiller, Mass Communication and American Empire (New York: Augustus M. Kelley, 1969), p. 80.

20. Ibid.

21. Ibid.

22. Dizard, op. cit., p. 125.

23. United States Information Agency Appropriations Authorization, Fiscal Year 1973, Hearings before the Committee on Foreign Relations, United States Senate, 92nd Congress, 2nd Sess. (March 20, 21 and 23, 1972), p. 328.

24. Quoted in Schiller, op. cit., p. 12.

25. Quoted in Gail Grynbaum, "United States Information Agency: Pushing the Big Lie," NACLA's Latin America and Empire Report 6, no. 7 (September 1972): 2.

26. Dizard, op. cit., p. 125.

27. Schiller, op. cit., p. 81.

28. Dizard, op. cit., p. 248; Barnouw, op. cit., p. 112.

29. Dizard, op. cit., p. 72.

30. Ibid., p. 99.

31. Ibid., p. 72.

32. Ibid., p. 248.

33. The Network Project, Domestic Communications Satellites, Notebook No. 1 (October 1972), p. 2.

34. American Broadcasting—Paramount Theatres, Inc., Annual Report 1961, p. 13.

35. Al Kroeger, "Merger Machine in High Gear," Television Magazine 24, no. 2 (July 1966): 42.

36. Ibid., p. 48.

37. Ibid., p. 42.

38. ABC-ITT Merger, Memorandum Opinion and Order, Dissenting Opinion of Commissioner Nicholas Johnson, 7 FCC 2d 287 (December 21, 1966).

39. Kroeger, op. cit., p. 48.

40. Quoted in 7 FCC 2d 296 (1966).

41. ABC-ITT Merger Case, Opinion and Order on Petition for Reconsideration, Dissenting Opinion of Commissioners Rober Bartley, Kenneth Cox and Nicholas Johnson, P & F Radio Reg. 2d 361 (June 22, 1967).

42. Quoted in 9 P & F Radio Reg.2d 57 (1966).

43. Quoted in 7 FCC 2d 295 (1966).

44. For a more detailed account of these episodes see 9 FCC 2d 593-599 (1967).

45. Quoted in Kroeger, op. cit., p. 46.

46. 9 P & F Radio Reg.2d 46 (1966).

6

THE CONGLOMERATE
COMPLEX

In 1955 the British broadcasting system was commercialized, culminating an extended campaign engineered by the London branch of the American J. Walter Thompson advertising agency. The following year an elated Canadian, Roy Thomson, was granted the television franchise for Scotland. Upon being awarded the franchise Thomson exclaimed that a television license is "like having a license to print your own money!" Since then Thomson's maxim has become legendary, as have the profits reaped by broadcasters.

In 1956 Chairman Emmanuel Celler of the House Antitrust Subcommittee interrogated George McConnaughey, then Chairman of the FCC. The following exchange occurred concerning a CBS-owned New York station:

Celler: In 1955, WCBS had a net income before federal income taxes of $9,375,339?

McConnaughey: Yes, sir.

Celler: And WCBS had a net investment in tangible broadcast properties as of December 31, 1954, of $409,484?

McConnaughey: That is correct, sir.

Celler: This means, does it not, that in 1955—and I give emphasis to these figures—WCBS recovered 2,290 per cent on its total investment in broadcast property?

McConnaughey: That's correct, sir.

Celler: You would say, would you not, those are high profits.

McConnaughey: Extremely high profits.[1]

Precise profit figures for an individual television station like WCBS are now carefully concealed from public view. Not eager to risk potentially embarrassing inquiries like those of Congressman Celler, the FCC has an agreement with broadcasters to keep secret from the public the specifics of broadcast profiteering. The only exceptions are the gross aggregated balance sheet data the FCC annually releases for all network operations combined, and all broadcast stations collectively. Consequently the revelations like that for WCBS are informational anomalies among available data on broadcasting finances.

But even the data doctored for public consumption, first by the broadcasters and then by the FCC, makes clear that "extremely high profits" are not financial anomalies. The profit picture that emerges from FCC statistics shows the most handsome profits concentrated at the top, in the networks. As the chief distributors of advertising and programming for the industry as a whole, the networks have the power to apportion the wealth. And that wealth is not shared equally. In 1970 the three networks sold $1.5 billion in advertising time for programs carried by their affiliates. The networks retained 87 percent of this amount, over $1.75 billion, and distributed the remainder. To the network-owned stations an average share of $2.5 million was allocated. The remaining 523 affiliates were left with shares totalling $394,000, on the average.

For the industry as a whole, broadcast revenues continue to rise annually. Revenues topped $3.7 billion in 1974, which is nearly a billion dollars more than the 1970 total. Profits have followed suit. In the 15-year period from 1960 to 1974, the ratio of pretax profits to depreciated tangible investments reached a startling high of over 200 percent in 1974. The low for this period was a healthy 53 percent in 1971.

Within the industry, the television networks and their 15 stations accounted for $213 million in 1972 pretax income. The other 475 VHF stations made $355 million, while the 173 UHF stations faced a collective deficit of $16 million. Clearly the networks fare best: their ratio of pretax income to depreciated tangible investments had never fallen below 100 percent in the 1960-1972 period.

Occasionally leaks develop in the walls of secrecy erected by the FCC and broadcasters, which provide enriching detail to the agregate evidence. Insiders at WNBC-TV in New York revealed that station's profit status for one year calculated on the basis of profits as a percent of gross revenues. For 1970, WNBC-TV's revenues tallied $50.6 million, of which 48 percent, or almost $25 million, was profit before taxes.[2] Reports of similar findings elsewhere suggest that such bonanzas are not limited to network-owned stations broadcasting in the largest television market. In fact, WNBC-TV's standing may be typical. The National Association of Broadcasters has reported that a sample of top 50 market stations showing profits in 1970 averaged from 30-38 percent return on revenues.3 And from far down in the market ranks comes the case of WISC-TV, the sole VHF broadcaster in the 110th-ranked Madison, Wisconsin market area. A vigorous citizen's investigation there revealed that WISC-TV retained 48 percent of its revenues as profits in 1967, and 40 percent in 1969. As one former station owner testified during Senate hearings in 1969, "Well-managed major market stations . . . are already earning 100 percent on invested capital. They will get their money back three times over in a three-year license period, and many have already made 10 and 20 times their licenses."4

Yet even FCC and NAB aggregate figures doubtless understate the affluence enjoyed by broadcast corporations. First, the concept of profit has been expanded in the world of modern corporate broadcasting to include benefits that do not show up on conventional profit-and-loss sheets. These include the opportunity to retain stand-by pools of employees, excessive reserves of equipment rarely used, and salary- and fringe-benefit arrangements for on-camera performers which resemble emoluments customary under the Hollywood star system. Nor are such hidden profits restricted to television entertainment productions. For example, an executive from WCBS-TV in New York lists the following annual "news" expenses for the station:5

A new set has been built in each of the last two years	$ 100,000
Anchorman Jim Jensen	150,000
A newly added anchorman	60,000
An announcer to say, "Here is Jim Jensen"	17,500
A fourth camera for an opening "beauty shot" from above (permanently mounted, and with no other use), including maintenance	52,000
Ten film crews, $125,000 each, including dar, two-way radio, and three men for each (five are superfluous, the executive said)	1,250,000
	$1,629,500

Secondly, profits are hidden through a variety of accounting devices and other practices employed in the service of obfuscation and deception. The favored technique involves passing on profits from a

station to other nonbroadcasting divisions of the station's parent corporation, thereby reducing the real profit levels for operations that
must be reported to the FCC. Thus stations often "rent" their studios
and their offices from corporate parents, and use in-house ad agencies to which they pay commissions far in excess of competitive rates.
This way the parent company retains the profits, reporting them as
broadcasting expenses, while segregating them as income from nonbroadcasting subsidiaries. WNBC-TV, for example, utilizes each of
these mechanisms and others, which has prompted one investigator
of that station's financial structure to estimate that an undoctored account would show the station profiting 70 percent on gross revenues.6
     The same expertise in accounting is applied to another set of
public reports, those required annually by the Securities and Exchange
Commission of all publicly traded stock corporations. While the SEC
regulations stipulate that each division of a publicly owned conglomerate must show its separate earnings, broadcasters typically avoid
the full disclosure intent by aggregating their soft-drink bottling operations, theater holdings and other unrelated interests with their broadcasting division. Even so, the SEC reports detail the standing of
broadcast profits comparative with profits from other types of commercial enterprises. The records of two diversified conglomerates
furnish examples. RCA, parent to NBC, reports that its broadcast
and related operations in 1970 accounted for 23 percent of the corporation's total revenues. But the profits contributed by broadcasting
the same year accounted for 50 percent of RCA's total. Similarly,
Westinghouse's broadcasting subsidiary collected 5 percent of the
corporation's revenues in 1970, but supplied the conglomerate with
20 percent of its total profits.
     SEC reports confirm the impressions of "extremely high prof-
its" that surface regardless of what source is cited or what profit
measure is used. But more importantly, they allude to the formative
role that these profits assume in generating a diversified, conglomerated corporate structure around broadcasting. The RCAs, West-
inghouses, and a number of less well known corporations have found
in broadcasting a profit center approaching indisposable proportions.
     If the regulatory climate allows profits to be not only uncon-
trolled, but even undisclosed, it also empowers the business owners
to use the telecommunications resources however they choose. Their
skillful use of it has shaped electronic communications into a mass
consumer delivery enterprise. The profit opportunities detailed above
demonstrate the enormous incentives, to business owners at least,
to make use of the communications resources in this way. But unre-
stricted profit opportunities have other consequences as well. The
accumulated income that is produced exceeds the owners' capacities

for earnings retention. Thus the initial business incentives generate others: to reinvest earnings through expansion and diversification.

In the previous chapter we examined one business expansion biography which exorbitant profits facilitated. In ABC's case international expansion was emphasized. Even so, within a decade of the company's creation it had also expanded domestically to acquire profit and production control over both the entertainment and informational programming it broadcast. The incentive to diversify led ABC into the ITT merger which was later blocked by legal developments. But other corporate broadcasters have been more successful in their diversification drives, which will be examined in this chapter.

The incentives which make expansion and diversification imperatives are acknowledged by broadcasters, but in a different vernacular. As the chairman of one station group explained, "Stations are considered milk cows by some broadcasters. They pull in the money from them while they move into other areas."[7] To some investment analysts, any commercial enterprise promising growth is a candidate for acquisition by diversifying broadcasters. But while there may be some truth in one analyst's claim that "broadcasters don't know what to do with all the money they've been making,"[8] corporate indifference to investment choices is clearly not one of the discernable outcomes of the diversification imperative. The investment patterns that do emerge can be differentiated into three forms, corresponding to the relative homogeneity of their economic lines: media conglomerates, concentric conglomerates, and diversified conglomerates.

MEDIA CONGLOMERATES

The term media conglomerates is given to broadcasting corporations which combine their television operations with newspaper, periodical and book publishing, radio broadcasting, and cable television to form horizontally integrated information empires.

A company like Cowles Communications, Inc., turned to broadcasting to rescue a stagnating magazine publication and was amply rewarded, fashioning an extensive media machine in book, periodical and newspaper publishing from its broadcast profits. Cowles began in 1936 with Look magazine and remained a single magazine publisher until building a radio-TV station in the media's early days. Adding to its broadcast holdings by acquiring a major market radio-TV station in 1963, Cowles began its expansion and diversification drive. By 1970 the company owned three radio-TV outlets, two in the top 75 markets, and over 45 subsidiaries engaged in publishing six mass circulation magazines, eight daily newspapers, 16 professional and

business periodicals, encyclopedias, directories and books. During
this time Cowles' newspaper acquisitions made the corporation the
eighth largest newspaper chain in the country. The corporation's dependence upon its broadcast profits, even with its numerous and diverse other properties, was reflected in its 1970 financial statements.
Revenues from broadcasting that year accounted for only 6 percent of
Cowles' consolidated revenues; but broadcasting accounted for 89 percent, or virtually all of Cowles' income. In 1971 Cowles transferred
several of its properties, including a TV station, to the New York
Times in exchange for 23 percent of the Times' stock. Several other
properties were sold at the same time so that Cowles now functions
as a management investment company with the Times' media holdings
being its most important asset.

Meredith Corporation, like Cowles, began in magazine publication, and for over 40 years was confined to publishing two monthly
periodicals. Its construction of a TV station in Syracuse in 1948
opened up growth opportunities which paid off rapidly: from 1951
through 1954, Meredith each year added other stations to its broadcast holdings. Beginning in 1959, these profits were turned elsewhere
as Meredith, over the next decade, purchased ten other corporations
engaged in newspaper publishing, elementary, high school and college
textbook publishing, consumer book publishing, and telefilm production.
Since 1970, Meredith has invested vigorously in microwave communications which provide ground linkage for satellite systems, and the
relay of information services nationwide. Through its interest in MCI
Communications, Meredith is part of an electronics consortium which
holds a coast-to-coast microwave monopoly reaching more than 6,000
miles through 41 states, with service areas covering 81 percent of the
nation's population. Still forming the profit core of its operations are
Meredith's five TV stations, all located in the top 75 markets, and its
six affiliated radio stations.

Cox Broadcasting, perhaps more than any other media conglomerate, has demonstrated how impressive gains in several media fields
can be achieved in less than a decade. The Cox philosophy of diversification was expressed by vice-president C. M. Kirtland in 1967:

We do know we'd like to bring into the home anything that's
going to come in, and we suspect that with broadcasting as
a base, there's a little bit of a head start because we're already in the home.[9]

Since that time Cox has departed occasionally from its own
version of home rule by indulging in automobile auction services and
technical publications in its portfolio of acquisitions. Its status in
home media services is, however, in keeping with Kirtland's ambition.

Cox's five television properties in the top 50 markets and its nine radio stations place it among the larger group broadcasters. Its daily newspaper holdings rank ninth among newspaper chains. And its CATV acquisitions combine to form one of the largest cable systems in the United States.

Cox Broadcasting Corporation was officially launched in 1964 when the four TV and eight radio stations owned by James M. Cox were consolidated under a common corporate title with Cox's seven daily newspapers. Diversification and expansion began promptly with three acquisitions in the next ten months, and 20 more have followed in the succeeding ten years. In the process Cox has added three newspapers; purchased or built cable systems in over 50 communities in 15 states, giving Cox enough franchises to make it number five in that industry; expanded its radio and TV holdings; purchased a substantial portion of the regional auto auctioneering services and facilities east of the Mississippi; and acquired magazine and directory publishing firms serving the commercial, military and aerospace electronics market.

While vigorously pursuing diversification, Cox has at the same time spared no money in its quest for additional major market broadcast properties, attesting to that medium's importance in the corporation's scheme for continued growth. Thus Cox's purchase of Pittsburgh TV station WIIC in 1964 for $20.5 million established a price record not broken until 1967. And its 1973 acquisition of KFI-AM radio in Los Angeles for $15.1 million exceeded by more than 30 percent the previous price record for any radio facility. Not surprisingly, the company's financial statements provide justification for these moves—while in 1970 65 percent of Cox's revenues came from broadcasting, fully 87 percent of the corporation's profits were supplied by that division.*

By and large, media conglomerates are formed by companies with established investments in earlier media of mass communications. (They are the crossowners discussed in Chapter 3, but in a majority of cases their newspapers are not located in the same cities as their television stations.) Upon securing a profit base in broadcasting, they have extended the investment logic which protected their original

---

*Cox's CATV interests are now separately incorporated as Cox Cable Communications, Inc., with Cox Broadcasting maintaining its controlling interest and interlocking directorate with the firm. Cox Broadcasting does not include its CATV interests in the financial reports it files with the SEC, from which the profit and revenue data reported here were taken.

print interests, and in some cases spelled their once impending stag-
nation. Thus developing media conglomerates have turned to investments in still newer technologies like cable and microwave; retrenched
their position in the older media of newspaper, periodical and book
publishing; and, at the same time solidified their broadcasting prominence through trading up to higher market, higher profit, stations.

CONCENTRIC CONGLOMERATES

Concentric conglomerates, on the other hand, represent a
younger generation of media enterprises first established on a broadcasting base. These are major corporations whose nonbroadcast interests are primarily in a single large enterprise, or group of closely
related companies—hence the term, concentric. Lacking a prior history of media experience, their diversification imperative follows no
common form, as in the case of media conglomerates. While there
are instances of investments into allied fields such as advertising and
entertainment, other acquisitions seemingly owe their logic only to
the personal predilections of their founding, and still reigning, fathers.
Thus these companies have been charged by some financial analysts
with indifference to the problem of selecting investments. For the
visibly incompatible corporate portfolios which sometimes result
suggest an indulgent and even capricious investment scheme which
only a profit-beseiged company could afford.

Concentric conglomerates in particular show an exceptionally
high ratio of profits to revenues contributions from their broadcasting divisions. A sample of major corporations in this category report, variously, that their broadcast divisions contribute only 20 percent to revenues, but 55 percent to profits; or 52 percent to revenues,
and 83 percent to profits; or 34 percent to revenues, and 252 percent
to profits. Wometco Enterprises, Capital Cities Broadcasting, and
Storer Broadcasting, respectively, are three such companies.

Wometco's broadcast profits have afforded it the financial
leverage to become one of the nation's major soft drink bottlers and
vending machine operators. In explaining Wometco's investment program, one industry executive argues that "a license for broadcasting
isn't unlike a franchise for a bottling company."[10] This would seem
to reflect the view that the only responsibility of local beverage or
communications distributors is to package the formula products of
national syndicates. Wometco also lists among its holdings the Miami
Seaquarium, 97 movie theaters, cable television systems in seven
states, and three wax museums.

Storer Broadcasting is a major group owner holding 13 broadcast properties along with extensive CATV interests in five states. Five of its six television stations are in the top 25 markets. Storer more than doubled its operating revenues in a single transaction when it acquired Northeast Airlines in 1965. This diversification in particular is referred to in industry circles as representative of the "hidden similarities between industries that make some seemingly wild moves a lot more reasonable," purportedly a reference to the fact that both broadcasting and air transport operations are government regulated.[11] The acquisition in this case has proved to be a continuing receptacle for broadcast profits, as Northeast's unbroken record of annual deficits occasionally topped the $20 million mark. In 1972 Storer merged the airlines with Delta, becoming a major Delta stockholder in the transaction. A new field for diversification was entered in 1973 when Storer purchased a Boston professional hockey team.

Like Storer, Capital Cities Broadcasting is a major group owner. It also owns a well-deserved reputation for station trading, with a remarkable record of 21 broadcast acquisitions and 12 station sales during the 1961-71 decade. It began the 1960s with four TV and three radio properties, and in 1972 owned six TV and 11 radio stations, managing to retain only one of its original TV properties. Amidst this shuffling of stations in the process of trading up, Capital Cities, like Storer, doubled its operating revenues through a single diversification move, when it acquired the merchandising publication operations of Fairchild Publishers in 1968.

These three companies and the subsidiaries they control each have annual operating revenues in the $100 million to $250 million range. The undisputed pacesetter among concentric conglomerates, however, is a company whose diversification record on the whole reflects a more visibly compatible portfolio—Metromedia.

The Metromedia success story, a favorite in the trade press, is an account of almost legendary growth from holdings of two TV properties with gross assets of $6 million in 1957, to assets of over $200 million and the status of a major group broadcaster and conglomerate in just 15 years. Like Capital Cities, Metromedia grew through trading up, first acquiring small market properties, then selling them off at handsome profits as major market stations became available for purchase. In 15 years Metromedia acquired 23 stations and disposed of eight, selling its small market TVs at profits ranging from 130 percent to more than 1000 percent. In 1974 it owned six TV stations, all in the top 25 markets, and 12 radio properties. In the course of these transactions, Metromedia simultaneously diversified, purchasing 33,000 billboards in 15 metropolitan

markets, franchises for sale of advertising space on 22,000 transit
vehicles in nine markets, and direct mail advertising lists totaling
over 52 million consumer names. As a result the corporation is now
the country's largest outdoor and transit advertiser, and the third
largest direct mailing firm.

Along the way Metromedia has also indulged in less conspicu-
ously compatible purchases, acquiring the Ice Capades and Mt. Wilson,
which it promptly transformed into an amusement park. These latter
investments have a subtle logic of their own, according to industry
analysts:*

> . . . when you have your own ready-built promotional and
> sales facilities, there is a method for pushing such seem-
> ingly capricious subsidiaries as the Ice Capades (Metro-
> media), Miami Seaquarium (Wometco) or Weeki Wachee
> Springs, Fla., resort (ABC).[12]

Thus the concentric gestalt which emerges from Metromedia's diverse subsidiaries might be labeled "coordinated marketing and promotion." As one company spokesman explains, "No matter how
round-about a route we take, in the end, we are probably the only
company in the country where all of our activities can be reported in
the advertising press every week."[13]

DIVERSIFIED CONGLOMERATES

To recapitulate briefly, there is a tendency among older firms,
already established in the printed mass media prior to the television
era, to use broadcast profits to fashion national media machines. Here
print, cable and broadcast communications are combined under one
corporate parent. In the case of younger companies first raised to
prominence on a broadcasting foundation, towering empires in service,
leisure and advertising industries are being wrested from the same
base. But the institutional integration of broadcasting with the industrial system as a whole is nowhere as advanced as in the case of the
diversified conglomerates grounded financially in electronic communications.

---

*Taft Broadcasting, another concentric conglomerate, also
owns multimillion dollar amusement resort centers in the South and
Midwest.

This probe of television's ownership structure began in order to consider the meaning of business control for the electronic communications flow. This required a look at incentives, which originated with profit opportunities, and as we saw, developed into imperatives of expansion and diversification. In the process the incentive structure grows more complex. For with multinational expansion and even more so with diversification, as we shall see, the control structures of broadcasting and those of the corporate industrial system are blurred. This encourages additional incentives to identify and explcit common interests between these two institutions. And with electronic communications conceived of and operated as a consumer delivery enterprise, there is a basic compatibility between the broadcasting business, and business in general. Opportunities constantly arise to use the consumer delivery enterprise to maximize shared interests in ways beyond the basic transaction of paid business access, which built these bridges in the first place. These opportunities are brought closer to home for the broadcasting corporations which expand, diversify, and become conglomerates.

There is another facet to consider, particularly so in the context of consolidating control structures, and in the context of the more cooperative incentives which appear to be developing. Do broadcasting's owners have the capacity to represent economically and politically diverse interests, needs and priorities? Have they the capacity to reflect faithfully political and economic antagonisms in those brief moments allocated for social intelligence purposes in a consumer delivery communications system? The capacity of the flow itself to service a diversity of interests is already undermined by its very shape. As we noted in analyzing business pluralism earlier, regular paid access privileges are awarded exclusively to business interests and commercial information is alloted more time than is social intelligence. The problem here, then, is to consider to what extent developing control structures and incentives facilitate diversity or frustrate it.

The development of a different kind of diversity in the control structure—one called conglomeration—is instructive. For example, ABC described its diversified interests in the language of a public relations release:

> When you go out to the movies in Tucson, you're watching ABC. Or in Chicago. Or Houston. Or Jacksonville. Because ABC owns the largest chain of motion picture theatres in the world.
>
> When you play a top ten record, you're watching ABC. Because ABC is one of the largest producers of records in the world.

When you're learning all about high-lysine corn
in Prairie Farmer, you're watching ABC. When you
ride in a glass-bottom boat or go out to see "Hell
in the Pacific," you're watching ABC. When you play
a top ten record or talk to the porpoises at Marine
World, you're watching ABC.
We're many companies, doing all kinds of entertaining things you probably didn't know we did. There's
a lot more to American Broadcasting Companies than
broadcasting.
We're not quite as simple as ABC.[14]

While ABC's insistence on its conglomerate status has grown
more vociferous since the ITT merger fumble, the diversification
enjoyed by the company's collegial superpowers is undisputed. RCA
Corporation, Columbia Broadcasting System, Inc., Avco Corporation,
Westinghouse Electric Corporation, General Tire and Rubber Company, and Kaiser Industries Corporation each exhibit a profoundly
extensive diversification through the industrial system. And in broadcasting, all are major group owners with four or more television
stations in the top 50 markets under their control.

Table 7 lists the most important telecommunications corporations in the United States, based on group ownership of at least four
TV stations in the top 50 markets in 1974. They are ranked according to the proportion of the national audience they have direct access
to. In actuality, several of these corporations have either network or
syndication systems which give them effective access to a greater
proportion of the audience than these figures suggest.

The diversified conglomerates on this list are led by the CBS
and NBC networks (the latter is a division of RCA), each of which
owns five TV stations located in cities which include 23 percent of
the national television audience. In addition, their network arrangements with affiliates give them virtually 100 percent penetration of
American TV homes. In 1974, CBS owned 14 radio stations and NBC
owned eight. The next ranking conglomerate, Kaiser Industries, owns
seven TV stations located in the top ten markets with access to 21
percent of the TV public. The General Tire and Rubber conglomerate
follows the networks with its TV stations broadcasting to more than
18 percent of the American public, along with a full complement of
radio properties. Next in order is Westinghouse, with five TV and
nine radio stations. Through its Group W division, Westinghouse is
also a major program supplier and syndicator to the rest of the industry. Avco, the remaining conglomerate on this list, with five TV
stations is primarily a regional, midwestern broadcaster, although
its syndicated programs are seen nationwide. Avco also held seven
radio properties in 1974.

TABLE 7

Access of Television Corporations to
National Audience, 1974

| Corporation | Stations Owned | | National TV Audience (percent) |
	TV	Radio	
Columbia Broadcasting System, Inc.	5	14	22.8
American Broadcasting Co., Inc.	5	12	22.8
RCA, Inc.	5	8	21.9
Kaiser Industries	7	3	21.3
Metromedia, Inc.	6	12	19.1
Gen. Tire & Rubber Co.	4	13	17.3
Spanish International Communications Corp.*	5	0	16.4
Wastinghouse Electric Corp.	5	9	10.7
Storer Broadcasting Co.	6	6	9.1
Taft Broadcasting Corp.	6	8	7.2
Capital Cities Broadcasting Corp.	6	11	7.1
Oklahoma Publishing Co.	6	2	6.6
Cox Broadcasting Corp.	5	9	6.5
Avco Corp.	5	7	3.9
Combined Communication Corp.	6	2	3.7

*All Spanish language stations.

Note: All group broadcasters owning four or more TV stations in the top fifty markets in April 1974 are included. The access percent is derived by summing the percentages of national TV households for each market where a corporation owns a TV station.

Source: Compiled by the author.

Each of the above, in addition to being among the largest television broadcasters in America, is also a well-diversified conglomerate corporation. A survey of precisely what these conglomerate operations entail suggests that their control structures, and the incentives which govern their decisions, are qualitatively different from the kind of mom-and-pop local broadcasting operation romanticized in the FCC's pluralist ideal. Conglomerated communications systems are vastly more complex operations. Their financial interests penetrate most of the industrial economy, and involve them with much of the government bureaucracy—federal, state and local. Even within their own house the complexity is profound. The management decisions within one subsidiary may have considerable effect on the fate of a separate subsidiary. This may sometimes produce internal conflicts within the conglomerate as a whole. At other times it may produce compelling opportunities for mutually advantageous actions and policies.

One thing conglomeration facilitates is a supramanagement authority overarching these separate enterprises. Its capacity to direct and coordinate would not exist lacking the financial and administrative bonds of conglomeration. That supramanagement may be faced with decisions of whether or not, and how, to exploit internal opportunities; of whether or not, and how, to resolve internal conflicts. The substance of these decisions invariably involves more than simply determining whether to use the liberal income of one subsidiary for the expansion of a different and perhaps struggling subsidiary.

But even that decision may place profit or revenue generating responsibilities on one enterprise which force it to minimize costs unnecessarily in terms of its own separate financial status. Conglomeration, however, probably makes separate and self-contained considerations something of a relic—a luxury of preconglomerated management structures.

The smooth and efficiently tuned bureaucracy, however, will endeavor to present as few problems as possible for the supramanagement to settle, whether they be conflicts or opportunities. At the higher levels of each subsidiary or group of companies, the management will likely be quite aware of the parent company's overall interests and how these must be integrated with their own self-interest in scaling the corporate hierarchy. Corporate socialization means management will anticipate opportunities for intraconglomerate cooperation and initiate action on its own. It means management will anticipate potentially severe conflicts, and thereby avoid actions and policies which might aggravate them. In other words, the efficiently managed and bureaucratized conglomerate will not have to be ordered by its high command to pursue certain policies and not to pursue others.

An appreciation of this is significant. This is because some observers of control systems conclude rather hastily that lacking specific executive orders to proceed here or retreat there, visible retreats and advances must not be related to the corporate environment (that is, the financial and administrative attachment of mass communications systems to the business conglomerate). The behavior of the business conglomerate, like its control and incentive structures, is at once more subtle and complex than that stance admits.

With these general considerations in mind, we will survey more specifically the diverse economic interests of communications conglomerates for insights into the substance of business incentives therein. To explore these diversified investments we will borrow the paradigm of the ABC public relations release.

When you're watching American entries in the international competition for space exploration, you're watching CBS, RCA, Westinghouse, General Tire and Avco.*

Each corporation is a profit beneficiary of the U.S. space program through contracts with the National Aeronautics and Space Administration. Much of the NASA space program is engineered by RCA, which provides logistical support for NASA's tracking stations around the world, as well as for the agency's space flight centers and its testing ranges in the United States and offshore areas.

Not limited to engineering phases, these corporations also figure importantly in hardware construction. RCA designs and builds Atmosphere Explorer satellites, and constructs major components of the Earth Resources Technology satellites. RCA and General Tire work jointly on the Nimbus weather satellite series. And many of NASA's satellite earth stations are being built by RCA.

The most expensive and also the most heavily televised NASA program, the Apollo moon shot series, is a major joint enterprise of the broadcast conglomerates. RCA provides ground support and engineering for the moon shots which are launched by General Tire's Apollo rocket engines, equipped with Avco's miniaturized communications equipment. Once in space, RCA's guidance systems control the course of the craft and maintain communications contact with ground control. Even on the moon the television system already resembles the mixture of corporate broadcasters earthside, with some lunar cameras designed by CBS and built by Westinghouse, and others developed by RCA.

With an eye on their future even as they showcase their latest hardware and software products to national television audiences, the conglomerates are preparing already for further space programs where they are assured a prominent role in design, construction and promotion. General Tire holds research and development contracts for NASA's proposed space shuttle. Avco is engaged in research and development on the Venus probe. And the profit prize, the Mars flight scheduled for the 1980s but still a political uncertainty, is

---

*Unless otherwise cited, the corporate activities described here and in subsequent sections are all referred to in the respective annual reports of these corporations for the period 1967-71.

presently being developed under NASA contracts with General Tire and Westinghouse (for the NERVA nuclear rocket engine) and RCA (for the companion Viking spacecraft).

The resolute determination with which the broadcast industry tirelessly presses space flight coverage upon its audiences is reflected in its own programming statistics. For example, during the Apollo 11 lunar landing in the summer of 1969, the hours of space coverage alloted by the networks collectively exceeded their grand total of television political interview programs, documentaries, news specials, news analyses and news roundups programmed during the entire general election campaigns of 1960, 1964 and 1968 combined.[15]

The strategic role of RCA in the space program is reflected by the fact that NBC invariably leads the networks in programming time alloted to space coverage. CBS, with few NASA contracts, consistently trails NBC in hours alloted to space missions. And ABC, with no contract incentives to spur its interest, barely devotes half the time typically set aside by NBC.

Saturation programming of the above sort continues relatively unabated, despite the fact that occasional viewer feedback suggests that audiences prefer the telefilm fantasy versions to the infinitely more expensive NASA productions. The industry trade organ notes, for example, that "as early as 1966, when the networks stayed with coverage of the emergency splashdown of the crippled Gemini 8, many viewers complained about pre-emption of regular programs," including one called Lost in Space.[16]

When America's proprietary health care system defends itself against consumer critics, you're watching CBS, Avco, General Tire, Westinghouse and Kaiser.

Like the medical profession itself, each corporation has concentrated its operations in a few specialized areas. But together these corporations input into every principal level of the health business, from the training of medical professionals, to the development of medical technology and hardware, to the development and administration of delivery institutions, and to the planning of health care systems for the society's future.

Medical education is the special province of CBS which, through its W. B. Saunders subsidiary, is a prominent publisher of medical books, journals and other materials for medical professionals.

The largest single area of expenditures in the American health system is allocated to hospital supplies and equipment, which includes the growing market for medical electronics.[17] For most of the broadcast conglomerates, entry into the medical equipment and electronics field has originated as a military spinoff from projects they first developed under defense contracts. Thus General Tire's Aerojet defense industries division manufactures sophisticated electronic

disease detectors, Avco develops and markets heart surgical technology, and CBS is under federal contract to apply its holography process to cancer and heart disease diagnostics.

At the delivery level, Kaiser Industries oversees, through its
financial and corporate interlocks with the Kaiser Plan, "the nation's
largest privately controlled prepaid health care system."[18] This
consists of 51 clinics and 22 hospitals located across six states, paid
for by more than 2 million subscribers. And in turn, the Kaiser empire of hospitals and clinics is a major client for Kaiser Industries.
"Kaiser engineers are the prime contractors for all Kaiser hospital
construction," and the Kaiser Steel, Kaiser Aluminum, and Kaiser
Cement and Gypsum companies market their respective materials to
the Kaiser Health Plan building program.[19]

Westinghouse has taken the lead in the area of health care planning, defining the nation's medical care problems in terms which conform with Westinghouse products, capabilities and resources. The
company's Medical Province department finds the basic crisis in
medicine to be "the increasing complexity and rising cost of health
care."[20] The Westinghouse-supplied solutions involve research,
computer systems, medical engineering, information handling, programmed teaching and advanced systems engineering. In this regard,
Westinghouse has designed a new generation of general hospitals
under a contract with the Department of Defense, and is analyzing
pediatrician efficiency for HEW.

When surveillance is promoted as the panacea for rising crime
and political disorder, you're watching Westinghouse, General Tire,
RCA, CBS and Avco.

With the provision of a laboratory for the development and testing of sophisticated surveillance technology in Vietnam and other
third world areas of American involvement, manufacturers of electronic snooping devices have been afforded a taxpayer-supported
opportunity to research and develop detection hardware for subsequent mass domestic deployment. Simultaneously, the wide publicity
given to crime wave statistics by the media, and the propagation of
law and order issues has been an undeniable aid to the development
of vast, new markets for this security technology among frightened
citizens and suddenly affluent police departments. Eager to capitalize on this newly cultivated consumer taste are the crime profiteers,
corporations which instinctively respond to social problems as potential markets for new technologies, in this instance surveillance
systems.

The involvement of the broadcasting conglomerates in this
field originates with the American military strategy in Vietnam. A
number of surveillance systems with instutional applications are
derived from Vietnam-inspired research. As examples, Westinghouse

has perfected radar pulsing equipment; RCA and CBS have concentrated on laser and television photographics; General Tire has specialized in infrared techniques; with Avco developing both infrared and radar systems.

One need only look to Westinghouse and CBS for evidence of what peacetime conversion of war industries can mean. CBS's Compass Link system of reconnaissance photography provides the kind of high resolution photographs of local comings and going of citizens which are apparently desired by police departments, given the demand reflected in its use in numerous American communities.[21] And oriented toward mass consumer security consciousness is Westinghouse's Security Systems Inc., which manufactures and markets an electronic home protection system that is a market leader. The complementary Westinghouse arsenal for the institutional security/espionage market is evidently the most diversified, including underground detection devices, ultrasensitive low light level imaging tubes, and a variety of surveillance-designed cameras.

When public education is beset with fiscal crises, when educators spar over test-centered learning, and when school administrators deflect pressures for community control by resorting to contract schools, you're watching Westinghouse, RCA and CBS.

On issues ranging from teaching methods to institutional control, these corporations are firmly committed to positions which channel the flow of educational funding into their hands. As producers of educational materials, they have been a vanguard in promoting the "materials teach" philosophy of learning, a revision of the older "teachers teach" maxim in favor during pre-Sputnik days when educational budgets were sparse. The materials that are currently alleged to do the teaching may be found in the Westinghouse, RCA and CBS warehouses: teaching machines, audio-visual equipment and materials, programmed instruction supplies, textbooks and tests.

A clue to the spiraling costs of public education may be inferred from some of the fashionable technologies these corporations are busy marketing to school systems prey to the notion that innovation is something you buy. The Westinghouse Learning Corporation, for example, promotes the use of computers and the application of operations research technology (developed for the military) to problems of classroom scheduling, facilities planning and student record-keeping. Apart from school operations, the company's Measurement Research Center functions in the area of instruction as a major processor of educational test results.

RCA has coordinated its testing and publishing operations to create the Criterion Reading Program, "which not only combines individual diagnostic testing of elementary school pupils with prescriptions for advancing their progress but also supplies the required

reading materials."[22] Such required materials are produced in con-
junction with RCA's extensive publishing division, which includes the
imprints of Random House, Alfred A. Knopf, Borzoi, Pantheon, Modern
Library, Vintage and Singer.

The CBS print and audio-visual instructional materials are produced by its Education and Publishing Group, which includes the Holt,
Rinehart & Winston publishing house. CBS notes with respect to its
subsidiaries in this division that growth and prosperity continue to
"depend largely upon the vagaries of public school funding," particularly the levels of state and federal aid to education.[23]

More ambitious than the marketing schemes for instructional
materials are the corporations' efforts to take responsibility for, and
control of, any number of the critical functions of public education
systems, including in some instances the entire school system itself.
This phenomenon, labeled contract schools, is an outgrowth of the
channelling of U.S. Office of Education funds in bulk to beleaguered
school districts. The districts, in turn, transfer under contract the
management of various programs, along with the funds, to corporations which take jurisdiction for a profit.

With its operations spread over five states, RCA stands as one
of the more aggressive marketers of school contracting arrangements.
In Florida, the State Department of Education has hired RCA to provide a staff development program for teachers in migrant worker
communities across the state. In Reading, Pa., the school system has
turned over its administrator and teacher development functions to
the company. In Texas, the Dallas Independent School District has
contracted with RCA for a Career Development Center where the
company provides counseling and career oriented curriculum plans
for several thousand high school students. Federal funds allocated
to the Camden, New Jersey schools began in 1970 to be channelled
to RCA, which has charge of a comprehensive program to upgrade
the local educational system. And in Delaware, RCA's Educational
Management Systems Group is under contract to develop legislation
for all phases of school construction in the state.

Such contractual arrangements have understandably provoked
contentious controversy, albeit limited to the circles of professional
educators. One can only speculate to what extent the containment of
such controversy may be related to the disinterest of the contractors'
broadcasting divisions in this dimension of the crises in education.
If television has paid selective attention to the fiscal dimension of
educational problems, it has favored a crisis which threatens to
stifle its newly found growth industry.

When the Department of Housing and Urban Development com-
mits itself to the unprecedented national objective of 26 million new
housing starts in the 1970s, you're watching Avco, Kaiser, General
Tire, Westinghouse, CBS and RCA.

Just a decade ago, no more than one of these corporations would have been considered part of the housing industry. That was before the creation of HUD, which removed the economic risk from low and moderate income housing. Now federally financed construction and an assortment of subsidies has turned housing shortages into new profit opportunities. Subsequently, Avco Community Developers was formed, now a major developer of master planned new towns and low income housing. RCA, through its Cushman and Wakefield real estate subsidiary, purchased Construction for Progress, a developer of low and moderate income housing. And CBS acquired the Klingbeil Co., a nation-wide developer and manager of residential communities.

Federal decisions in the early 1970s expanding existing programs to include mobile homes among low income housing projects (a veiled retreat from construction goals announced earlier) represented a victory for mobile home industries like Avco and General Tire. The latter is a major manufacturer of mobile home accessories and interiors, while Avco builds and manages mobile home parks.

Leadership in following the federal dollar is shared by Kaiser (the only one of these conglomerates previously a part of the housing industry) and Westinghouse. Like Kaiser Cement and Gypsum, a major wallboard manufacturer, the fortunes of the Kaiser group, including Kaiser Sand and Gravel and Kaiser Steel, have historically been dependent primarily upon the level of new housing starts. Consequently, Kaiser has been a prime contractor with HUD for inner city housing programs. At the national level, Kaiser's responsibility has ranged over all the major program phases, including housing design, project supervision, and the responsibility for developing federal plans. Locally, Kaiser is building federally financed low and moderate income housing from Florida to California. In San Jose, for example, Kaiser has been selected to ascertain and define the housing needs of the poor, and to develop a master plan to meet them.

Just months after HUD was created by an act of Congress, Westinghouse launched a major initiative in pursuit of federal housing funding, purchasing a manufacturer of built-in home furniture, acquiring a series of local and regional construction firms across the country, and creating two new subsidiaries to manage the company's construction programs. As Westinghouse resports, "For the $20 billion a year home building market, the Company has formed the Urban Development Coordinating Committee, a 'think tank' with representation from all divisions of the Company," including, presumably, the broadcasting division.[24] Simultaneously, Westinghouse created the Urban Systems Development Corporation "to develop, build and sell low income housing through government sponsored housing programs," and to carry out strategy developed by its think tank.[25] The

USDC is conveniently headquartered just a short distance from the
HUD offices, in Arlington, Va. Its first project was a study conducted
for HUD of the "complex construction problems of low income housing" in 25 different American cities.[26] By 1971, barely three years
later, USDC was itself building housing units in 26 different American
cities, an undeniable tribute to the systems approach evidently utilized
by Westinghouse.

When the federal government launches innovative social programs in a War on Poverty, you're watching Avco, Westinghouse and
RCA.

Participation of the poor in the planning and administration of
antipoverty programs has received some attention in the media,
enough so as to render that participation controversial. On the other
hand, corporate participation has proceeded virtually unmentioned.
A review of the antipoverty grants to several broadcasting conglomerates—at a profit—suggests that they may have secured a more last-
ing presence in these programs than was ever achieved by client
groups.

Avco, for example, has operated Office of Economic Opportu-
nity-Labor Department Job Corps centers in the states of Maine and
Washington. RCA ran the Job Corps center in New York City, and
one for women in Pennsylvania, while Westinghouse has operated an
Indiana center.

A related project, funded by the Labor Department, is the Concentrated Employment Program (CEP). Westinghouse has been a
prime recipient of CEP funds for the establishment of Remedial Education and Job Development Centers in Manchester, N.H., East St.
Louis, Ill., and Washington, D.C. RCA operates the Choanoke Area
Development Center for migrant farm workers in North Carolina.

Under the Model Cities program, Westinghouse has found yet
another market for its Systems Operation department. This time in
Baltimore, Westinghouse established the federally financed Model
Urban Neighborhood Demonstration, a project which uses an engineering approach to encourage 20,000 ghetto dwellers to find solutions to their housing problems.

Corporate participation and control continues in a variety of
other programs as well. Under the Manpower Development and Training Act, RCA is paid to train the hard-core unemployed in the major
urban centers of the nation, including New York, Chicago, Los
Angeles, and in Camden, N.J. (where, as was noted previously, RCA
holds a comprehensive school contract). Some training of Vista and
Peace Corps volunteers has been contracted out to Westinghouse.
Publicly funded extracurricular and educational programs are operated by RCA at the Cornwells Heights Youth Development Center in
Pennsylvania. The only government funded national evaluation of the

Head Start program was conducted by Westinghouse. And conglomerate
Avco was the first company selected for participation in a government
funded training program called the Federal Test Program for Job De-
velopment. The terms of this grant enable Avco to be paid to hire
Roxbury ghetto dwellers for its new printing diversification venture in
Boston.

When America goes off to war in Vietnam, Laos, Cambodia and
Thailand, you're watching Avco, CBS, Kaiser, General Tire, RCA and
Westinghouse.[27]

With the exception of CBS and Kaiser, all of the above broadcasting conglomerates appeared on the list of top 100 military contractors for fiscal year 1972, a status Avco, General Tire, RCA and
Westinghouse shared every year in the preceding decade. A full accounting of the services supplied by these corporations, and paid for
by the military, could potentially fill volumes. For their products and
expertise have been involved, directly or indirectly, in every phase
of military policy for a decade. And in some cases the partnership
extends back over generations of wars, cold and hot. These conglomerates, in short, represent the corporate face of the military-industrial complex.

When U.S. troops and weapons embark for foreign shores, these
corporate trademarks carry them to their destination: if by air, Avco;
if by surface ship, Kaiser; if by submarine, Westinghouse. Avco, which
calls itself the nation's leading subcontractor for aircraft structures,
is also one of the nation's leading beneficiaries of the Indochina war.
In 1966, the year of the major American buildup, its military contracts
quickly doubled, to an annual total of more than $500 million. Troops,
ammunition, vehicles and other cargo are flown to Indochina literally
on the wings of Avco. The company's Aerostructures division produces the structural components, including the wings, for the military's C-130 Hercules and the controversial C5 series of air trans-
ports.

When tanks are deployed on foreign shores, their delivery comes
via Kaiser's new generation of LSTs, or tank landing ships, engineered and produced by the company's shipbuilding subsidiary, NASSCO.
Transportation on land for the military is also provided by Kaiser
through its jeep division. From 1964 to 1968, Kaiser manufactured
more than 100,000 cargo and ambulance trucks for the military for
the sales price of more than $1 billion, providing its jeep subsidiary
with record profits first in 1967 and again in 1968.* Like Avco, Kaiser

_____

*In 1970, Kaiser transferred its jeep division to the American
Motors Corp. in exchange for controlling stock interests in that company.

Industries as a whole benefitted handsomely from the military buildup
in 1966, more than doubling its defense contract totals from the previous year. And with two cement divisions and one fluorspar processing plant in Thailand, the American presence in Indochina provides
Kaiser with another kind of overseas insurance in addition to the financial coverage it has acquired from civilian agencies of the government.

The Westinghouse atomic fleet, operated by the U.S. Navy, is
that company's most recent product in a continuing line of naval craft
and armaments. Since 1942, Westinghouse has pioneered 13 generations of torpedoes, including the first to carry a nuclear warhead. In
torpedo development and design alone, Westinghouse has produced
over the years twice as many models as its nearest competitor. Fruits
of the company's systems approach, which it applies to marketing and
promotion as well as to product development, are reflected in Westinghouse's systematic integration into naval policy and hardware at
every strategic level. Midshipmen are first introduced to the company through leadership training courses prepared by Westinghouse
for the U.S. Naval Academy. There aspiring midshipmen learn that
successful naval careers are crowned with lucrative Westinghouse
retirement plans. In fact, so many retired admirals and other high-
ranking naval officers have joined the executive payrolls of Westinghouse that the company has been singled out by Senator William
Proxmire for excessive practices of this kind.[28]

Still, the tribute exacted for such favors is impressive and
compelling. For example, the Atomic Energy Commission's Bettis
Atomic Power Laboratory is virtually indistinguishable from what
would otherwise be called a wholly owned Westinghouse subsidiary.
While Bettis is nominally a facility of the AEC, the laboratory is
operated rent-free by Westinghouse for the AEC.

Parenthetically, these privileged arrangements and others like
them are in fact routine practices for Westinghouse, and have been
for some time. Beginning in the 1950s, and repeatedly since then,
the company has been cited in Government Accounting Office audits
of defense contractors for excessive costs, unreasonably high profits,
interest-free use of government funds, and rent-free use of government property.[29]

The majority of the Navy's nuclear fleet is powered by Westinghouse reactors designed and developed at Bettis.* This includes the

---

*Research at Bettis has also facilitated the company's position
as the leading manufacturer of nuclear power plants throughout the
world.

entire Polaris-firing submarine fleet, the majority of nuclear surface ships, and the launching equipment for all submarine-fired Poseidon ICBMs. With an assortment of other military hardware in both planning and manufacturing stages, Westinghouse is assured a continuing flow of military funds in the future at at least the level (10 to 12 percent of total company sales) that it experienced during the decade of the Vietnam War.

On the ground and in the air over Southeast Asia, the war proceeded in conformity with the capability of products designed by these conglomerates. Aerial bombing is facilitated by Kaiser's production of electronic optical viewing systems for B-52 bombers refitted for Vietnam deployment. Kaiser also designs and installs the radar and avionics systems on a variety of all-weather attack aircraft. And the Navy F-14 fighter, controversial because of cost overruns and crash-proneness, is supplied with avionics systems by Kaiser.

At lower altitudes the war was conducted by Avco-powered helicopters, including both gunship and cargo varieties. Avco manufactures 80 percent of all U.S. helicopter engines. Inside the aircraft, RCA's interior voice communications systems aid the monitoring processes of the helicopter-borne tactical radar, the latter being a Westinghouse concept. The radar permits unparalleled detection of trucks, tanks and other aircraft, and provides a complementary system to Westinghouse's photographic quality side-looking radar used aboard surveillance planes in Vietnam.

Filling out the array of observational technologies is the RCA-produced system of satellites, and the company's line of hand-held tactical radar. Together these systems are used to coordinate and direct napalm strikes, artillery barrages, or a rain of what Avco calls its ''antipersonnel kill mechanisms.'' The ordering and direction of such assaults by armchair generals situated in Washington, and the commandeering of the war by the executive branch of the government alike, is made possible by CBS's Compass Link system. The system combines television and satellite technologies to relay high resolution aerial photographs of ground activity in Vietnam to Washington only moments after the photos have been taken.*

---

*In 1973, a U.S. Congressman and former staff member of the President's Council of Economic Advisors stumbled across the first of the CBS Compass Link projects. His further investigations showed that with the withdrawal of American ground forces, CBS had conceived of a more elaborate Compass Link system. This project was designed to obviate conventional Air Force communications and enable Pentagon generals to issue bombing instructions directly to pilots after

The antipersonnel weapons used in Vietnam, considered criminal by the International War Crimes Tribunal in 1967,[30] are designed and manufactured by Avco and General Tire. Antipersonnel weapons, it has been noted,

> have no effect against buildings or factories; they can do very little damage to military fortifications or vehicles; in fact, they cannot even harm military equipment, targets or personnel protected behind sandbags. They are aimed at only one target—people. . . [31]

Antipersonnel bombs have elsewhere been described as "the deadliest weapons being used against people in Vietnam,"[32] and as having the killing power approaching that of small, tactical nuclear weapons.[33] The military name for these weapons is cluster bombs. Most are dropped by aircraft; some are activated by land mines. The bombs themselves are designed to house a diversity of weapons packages, including flachettes, beehive darts, steel and gravel pellets, incendiaries like napalm and white phosphorous, and biological weapons. General Tire holds the distinction of being one of only two American companies which manufacture entire weapons systems. Between 1966 and 1969 General Tire sought and acquired 19 different contracts for the production of such weapons. Another supplier, Avco, has achieved special prominence with its sophisticated innovation known as the "jungle canopy penetration fuse." The latter was developed specifically for use against human beings in Vietnam. The company also assists the Air Force with chemical and biological warfare research.[34]

Elsewhere in Vietnam, RCA is active through an NBC International contract to supply managerial, technical and engineering serv-

---

receipt of aerial destruction photographs. This redesigned two-way system is of obvious strategic importance to the military's ability to continue bombings even in the absence of ground troop support. Congressman Les Aspin said this implicated CBS in the Asian bombing and called for the company to "get out of the war business now," because it was "potentially compromising" to its news operations. A CBS spokesman defended the company by claiming CBS was not responsible for what the military did with its ideas and equipment. While the public exchange was reported in some newspapers, CBS News declined to carry a story on it. (Capital Times, Madison, Wis. [June 7, 1973], p. 13.)

ices to the South Vietnamese ministry of information.[35] The precise
range and extent of the company's services has never been clearly
defined. It is known that during the major American buildup of the
war, NBC's coordinator for international broadcasting management
services functioned as the director of South Vietnam's television network.[36] Whatever its dimensions, this project has been coordinated
by the Joint U.S. Public Affairs Office (JUSPAO) in Saigon, the propaganda and psychological warfare arm of the American military in
Southeast Asia. According to one source, JUSPAO has been delegated
responsibility for martialing "American support for the war by favor-
ably influencing newsmen."[37]

RCA's participation in military propaganda and psychological
warfare actions is scarcely novel. RCA's partnership with the military through hardware, services and personnel dates back to the company's founding at the end of World War I. At that time it was the
naval branch of the military that was most dependent upon wireless
communications for tactical operations. Because the young wireless
communications industry was dominated in the United States by a foreign firm, British Marconi, the Navy engineered its expropriation in
1919, and supervised the creation of a quasi-government corporation
known as the Radio Corporation of America. For control of the company, ownership was vested to General Electric, Westinghouse and
American Telephone and Telegraph, with provisions included to provide for Navy representation on the board of directors. For its em-
bryonic period, a general was selected as the company's president
to consolidate military ties.

When commercial broadcasting proved to be a profit bonanza,
RCA bought out its original corporate investors. But its military ties
were not relaxed. In 1936 a military engineer was named NBC president, and his army associates were subsequently appointed to NBC
managerial posts.[38] With the coming of World War II, NBC was found
to be compliant with military objectives in numerous respects, including the production of regular programs aired over its commercial
affiliates, such as the Army Hour, Labor for Victory Series, and Victory at Sea.[39]

Following World War II, Brigadier General David Sarnoff re-
placed retiring General James Harbord as RCA board chairman and
president. During the first Eisenhower administration, Sarnoff retained for himself the position of NBC president, in addition to his
other corporate duties, and Sarnoff used his position to shepherd RCA
and NBC into an ever closer embrace with the military, a policy he
continued until his retirement in 1970. During the Eisenhower years,
Sarnoff promoted the military-political uses of RCA's technology,

from its equipment to its propaganda services. At Fort Meade, Maryland, Sarnoff demonstrated to fellow officers how future battles could be coordinated by TV, a precursor of the electronic battlefield. He served as an advisor for the CIA's Radio Free Europe and Radio Liberation. In consultation with the CIA, the U.S. Information Agency, President Eisenhower and Nelson Rockefeller (then Special Assistant to the President on psychological warfare), he developed in 1955 a comprehensive strategy document titled, "Program for a Political Offensive Against Communism."[40] Known as the Sarnoff Plan, it urged expanding American communications control abroad in noncommunist countries for propaganda purposes; penetrating communist countries through aircraft drops of transistor radios designed to receive only American transmitters; disseminating intensively the cold war message to the American public via radio and TV; and establishing special political-psychological warfare schools to train communications propagandists. Sarnoff managed to personally implement most of his proposals. For example, the CIA's Radio Liberation propagandists were trained at NBC studios, NBC International took charge of the South Vietnamese communications system, and the transistor bombing plan was acted upon by USIA-JUSPAO in Vietnam.[41]

Evidence suggests that the military-communications conglomerate interface will persist, even with the winding down of the war in Indochina. For the advent of the 1970s was marked by accelerated funding for strategic weapons production. The Navy awarded RCA its largest contract in a decade (ultimate value in excess of $1 billion) for production of the Aegis missile system, a new generation of guided defensive missiles. Increased ABM funding assured Avco of contracts for the company's ABM radar system, and the firm entertained hopes for a revival of SST aircraft construction in order to reclaim its status as the principal SST subcontractor. In 1971, the Navy launched its campaign for a new fleet, claiming Soviet naval supremacy, and letting significant new ship building contracts to Kaiser. And the automated battlefield system, estimated to cost $20 billion if funded, had moved beyond the concept and design phase, assuring all the conglomerates of substantial production contracts, particularly RCA and Westinghouse.

The preceding inventory of conglomerate investments is intended to be suggestive, not exhaustive. It gives evidence to how extensively the electronic communications system is simply a business adjunct to corporate America. The further significance of this for the shape of the communications flow will be considered in the final chapter.

NOTES

1. A lengthy excerpt of this interrogation is reprinted in Harry J. Skornia and Jack William Kitson, Problems and Controversies in Television and Radio (Palo Alto: Pacific Books, 1968), pp. 53-58.

2. Bill Greeley, "The World's Richest TV Station," More (January 1972): 5.

3. Broadcasting (August 9, 1971), pp. 24-25.

4. Testimony of Anthony Martin-Trigona, Amend Communications Act of 1934. Hearings before the Communications Subcommittee of the Committee on Commerce, U.S. Senate, on S.2004 (December 1-5, 1969), Part 2, p. 502.

5. Ron Powers and Jerrold Oppenheim, "Is TV Too Profitable?" Columbia Journalism Review (May/June 1972): 12.

6. Greeley, op. cit., p. 6.

7. A statement of Hazard Reeves, chairman of Reeves Telecom, quoted by Walter Spencer, "Why Diversification is the Name of the Game," Television 24, no. 10 (October 1967): 41.

8. This opinion was offered by Emmanuel Gerard of Roth, Gerard and Co., a New York brokerage firm with close ties to the broadcasting industry; ibid.

9. Ibid., p. 60.

10. Ibid., p. 43.

11. Broadcasting (January 24, 1966), p. 31.

12. Spencer, op. cit., p. 43.

13. Ibid.

14. Adapted from a series of ABC advertisements appearing in Television Factbook 1969-70 39, vol. 1. (Washington, D.C.: Television Digest, Inc., 1970).

15. This results from comparing statistics reported in Broadcasting (December 11, 1972), pp. 15-16, with others reported in Malachi C. Topping and Lawrence W. Lichty, "Political Programs on National Television Networks: 1968," Journal of Broadcasting 15:2 (Spring 1971): 176.

16. Broadcasting (December 11, 1972), p. 15.

17. Health Policy Advisory Center, The American Health Empire (New York: Vintage, 1970), p. 95.

18. Judith Milgrom Carnoy, "Kaiser: You Pay Your Money and You Take Your Changes," Ramparts 9, no. 5 (November 1970): 28.

19. Ibid., p. 30.

20. Westinghouse Electric Corp., 1968 Annual Report, p. 14.

21. The Network Project, Domestic Communications Satellites, Notebook Number One (New York: October 1972), p. 15.

22. RCA Corp., 1971 Annual Report, p. 22.

23. Columbia Broadcasting System, Inc., 1969 Annual Report, p. 5.

24. Westinghouse Electric Corp., 1968 Annual Report, p. 6.

25. Ibid.

26. Ibid.

27. Unless otherwise cited, the corporate activities described in this section are referred to either in the annual reports of these corporations, 1967-71, or in the company profiles available in "Corporate Military Contracting 1971," Economic Priorities Report 2:4 (January/February 1972).

28. Congressional Record (March 24, 1969), p. s3072.

29. Richard F. Kaufman, The War Profiteers (Indianapolis: Bobbs-Merrill, 1970), pp. 27, 114-15.

30. See Jean-Paul Sartre and Arlette El Kaim-Sartre, On Genocide and a Summary of the Evidence and the Judgments of the International War Crimes Tribunal (Boston: Beacon Press, 1968).

31. The Council on Economic Priorities, Efficiency in Death: The Manufacturers of Anti-Personnel Weapons (New York: Perennial Library, 1970), p. ix.

32. Quoted in Efficiency in Death . . ., from Frank Harvey, Air War Vietnam (New York: Bantam, 1967), p. 5.

33. John S. Tompkins, The Weapons of World War III (Garden City, New York: Doubleday, 1966), p. 116.

34. National Action/Research on the Military Industrial Complex, Weapons for Counterinsurgency (Philadelphia: American Friends Service Committee, 1970), p. 28.

35. Ralph Tyler, "Television Around the World," Television 23, no. 10 (October 1966): 59; "NBC Agrees to Help South Vietnam Build a TV Network," Wall Street Journal (July 1, 1966): 18.

36. Tyler, op. cit., p. 59.

37. Gail Grynbaum, "United States Information Agency: Pushing the Big Lie," NACLA's Latin America and Empire Report 6, no. 7 (September 1972): 6-7.

38. Erik Barnouw, The Golden Web (New York: Oxford University Press, 1968), p. 125.

39. Ibid., pp. 162-63, 230; and Erik Barnouw, The Image Empire (New York: Oxford University Press, 1970), p. 14.

40. Broadcasting (May 16, 1955), p. 114.

41. Barnouw, The Image Empire, op. cit., p. 145.

Surveying the diversified interests of communications conglomerates makes at least one thing certain—they know a great deal more about housing, health care, poverty, education, surveillance and military affairs than one would ever guess watching their television screens. In one sense that may be just as well. For the business viewpoint on the issues of health care, military intervention abroad, public education, and so on is understandably self-interested, just as the business viewpoint on how to use national communications resources is self-interested. What we see in the case of diversified conglomerates is a few mammouth organizations actively shaping military policies and implementing them; actively shaping social welfare, health and education programs and implementing them; producing a multitude of consumer items and shaping consumer interest in them; and controlling the substance and shape of the communications flow to the public on all these matters of public interest.

It appears that vast portions of the industrial system have been ransomed to the telecommunications monopolies. Or perhaps some of the prime engines of that system, understandably eager to acquire unparalleled profit centers, have simply added these monopolies to their burgeoning portfolios. In either case their control structures have become fused. The point is that those who have the power to decide which, if any, issues are to be discussed, and how, and when, and by whom, have more interest in these matters than just the financial prosperity of their communications operations. They have a stake in these issues themselves. For that reason, a communications flow which is long on sports, telefilms and variety shows; saturated with commercial messages; punctuated by headline news; and mostly silent on public affairs and the opinions of people who happen not to be business and government elites—perhaps that kind of communications flow is preferable to one wherein the conglomerates use their information resources more aggressively in behalf of their own interests.

This suggests that there may be good reason not to encourage any more informational and public affairs programming than is presently provided, as long as business conglomerates remain in control of the communications system. A shift to policies which would virtually compel such programming increases is gaining favor at the Federal Communications Commission, as we noted earlier. However, this paradox demands a more careful examination of the control and incentive structures that communications conglomerates create.

Surveying the conglomerated business structures which enmesh the electronic communications system, we can hypothesize two contending sets of incentives which would impinge upon broadcast subsidiaries. The first of these we have already referred to—the incentive to exploit intraconglomerate opportunities and avoid intraconglomerate conflicts in the shape and substance of the communications flow itself. In Chapter 6, a range of possible interest opportunities and conflicts of interest was described. Before considering the practical effects of these incentives, we must note a second possibility.

Incentives may be generated simply by anticipation of the imperiled legitimacy facing a system of public information which is chiefly accountable only to business interests, and more specifically, conglomerate businesses. Owners may desire to insulate their communications businesses from the rest of the corporate family. They may choose to proceed by enhancing what is called the professional approach to information. The significance of professionalism as a countervailing incentive will be appraised after first examining what it is that professionalism is supposed to restrain.

A BUSINESS INFORMATION ARCHITECTURE

In the American system, the very idea of using communications resources for profit has assumed one thing—the industrial system's need to manage consumer demand. The availability of mass communications resources like broadcasting causes unprecedented opportunities for such management. And the health and survival of profit-based broadcasting is made dependent upon how fully its resources for consumer management are exploited by industry. What this means over time is that the more the production of goods and services is adapted to this consumer management instrument, the more financially prosperous the business of communications itself becomes, and the more critically dependent the entire industrial system is upon it.

The effects of this on communications control and incentive structures have already been noted, but deserve a brief review here.

Business owners shape the communications flow to facilitate consumer management for other business interests. The owners reserve for other businesses the exclusive privilege of regular paid access. Thus a predominant feature of the flow, consumer delivery, is fundamentally based on the incentive to exploit an intrabusiness opportunity for cooperation.

The shape of the communications flow also suggests action on another fundamental incentive—to avoid business conflicts. First, exclusive business access means that potentially antagonistic voices can be excluded altogether, or carefully selected for exposure. Second, the allocation of most broadcast time to entertainment is openly justified as the most efficient lure for audience exposure to business messages. We noted before how an appreciation of this affected the spread of telefilms and an increase in paid business access.

Thus the general shape of communications under business control is evidence that incentives to exploit intrabusiness opportunities and avoid business conflicts are too compelling for the owners to ignore. The history of expansion abroad shows that the advantages of intrabusiness coordination are not only recognized, but exercised, and promoted by some agencies of the government. The phenomenon of broadcaster diversification is, after all, the product of an enormous transfer of wealth from manufacturers of consumer goods to broadcasters in exchange for exclusive access rights and compatible programming. And diversification signifies a tendency to fuse business control structures, thereby increasing incentives for intrabusiness coordination and cooperation.

We must ask ourselves at this point, whether the conglomerate, as the most modern and sophisticated form of business control, can improve upon the incentive structure that already exists; an incentive structure so fully exploited, that it fostered the development of the communications conglomerate itself. Can the incentives to exercise intrabusiness opportunities and avoid conflicts with business interests be enhanced by rearranging them as intraconglomerate opportunities and interests?

Before proceeding any further, it must be noted that it is too soon yet to fully appraise the ultimate effects of conglomerated control structures. This form of business control is still too new for its consequences to be decidedly manifest. While some broadcasters, like RCA, have been vested with business diversity for some time, others, like CBS, have only recently acted to diversify. And real conglomeration, like the economic use of the term itself, is not quite a decade old for these communications owners. Thus while the broadcasting subsidiaries are long accustomed to the framework of incentives which business brings to communications, their present personnel are no doubt equally habituated to expecting a separation of business powers

that was in accord with a separation of business control, before conglomeration. Not until a second generation of executives is installed,
one more acclimated to the distinctive managerial conditions of conglomeration, will the potentials of this structure be realized.

To put it another way, conglomeration is too recent a phenomenon for very many communications specialists to have sought employment with the hope that they may, like the worker who rises to foreman
and eventually factory president, climb from program director to network president, to chief executive of a multinational diversified enterprise.

The owners themselves are not ignorant of this fact. Recent executive shuffles at CBS, for example, indicate an impatient concern
with speeding up that company's adaptation to the new realities of corporate conglomeration. Its president since 1946, Frank Stanton, came
to the company with a Ph.D. in psychology when the CBS business was
radio broadcasting. Stanton ascended the corporate ranks as a research
specialist. In replacing Stanton in 1971, CBS didn't turn to any of his
underlings, groomed and poised for the presidency. Instead its directors sought executives more practiced and comfortable with managing
conglomerates. First, Charles T. Ireland, the senior vice-president
of ITT, was named CBS president.* After Ireland's death in 1972,
Arthur R. Taylor, the senior vice-president of International Paper,
was chosen as Ireland's successor.

But if the impact of conglomeration is still tentative, there are
nevertheless signs of new opportunities and incentives which certify
its presence.

The prominent and lucrative role the conglomerates enjoy in the
nation's space exploration program was described in the previous
chapter. There it was noted that the networks had allocated more
broadcast time to a single space adventure than they had to coverage
of national electoral campaigns spanning three presidencies and a
decade. As this example suggests, the electronic information flow is
expanded in some areas of obvious public concern, and relatively con-

---

*The CBS decision to move outside of its own hierarchy in
search of top command personnel was evidently undertaken with great
care. After being thoroughly scrutinized by CBS, Ireland reported
that "the job they did on me far exceeded anything the CIA did in clear-
ing me for my ITT work." Ireland was characterized in the industry
trade press as one who "represents a new kind of business thinking,
one that may at first appeal more to investors than to communications
traditionalists."

tracted in other areas. However, just as important is the fact that the area of expansion corresponds to areas of interest and priorities of commonly owned businesses. Here conglomeration extends new incentives: the opportunity to serve business needs for continued public support and federal funding of NASA contracts.

Evidence of similar incentives, inferred from instances where intraconglomerate business opportunities have been exploited, can be cited in areas besides space programming, and for conglomerates other than CBS and RCA.

Consider, for example, a major documentary telecast over Westinghouse-owned stations, and others, described as follows in a full-page ad taken out in the industry's trade press:

> There's one problem our cities are facing which make all the others obsolete.
>
> New Yorkers ride to work in subway trains built 35 years ago.
>
> Boston's transportation system is tottering on the brink of collapse.
>
> . . .
>
> The fact is, the public transportation systems in our cities are such a mess that in a few short years they may actually prevent people coming to the cities to live and work.
>
> In a one hour special, "The Battle for Urban Mobility," Group W illustrates the struggle to find workable systems of transportation in urban areas.
>
> Is the answer more highways? Modernized forms of public transportation? Or both?
>
> "The Battle for Urban Mobility," narrated by Rod MacLeish and produced by Group W's Urban America Unit, makes one thing painfully clear: If we don't do something to make it easier for people to get in and out of our cities, someday it will all just stop. First the people. Then the cities.[1]

However, what is not made painfully clear to viewers is that the Westinghouse Electric Corporation, corporate parent to WBC (the Group W broadcasting subsidiary), is the major supplier of urban mass transportation systems in the United States. In 1967, Westinghouse estimated that in the following decade there would develop a mass transit products market worth $11 billion in revenues, provided that 45 urban areas decided to build mass transit systems from scratch.[2] When a third of that decade had already passed, only a handful of cities had invested in mass transit systems. Hence, Westinghouse began waging its battle for the urban mobility market, a market which, to the company's disappointment, had failed to materialize. In time for the documentary filming, however, those "New York subway trains built 35 years ago" were being supplemented by the arrival of modern electric trains manufactured by Westinghouse. And the Boston system, "tottering on the brink of collapse," was being rejuvenated by a new supply of rapid transit cars manufactured by Westinghouse. And for the nation's first new rapid transit system in 60 years, the Bay Area Rapid Transit (which cost more than 1 billion in construction), Westinghouse was providing virtually all of the propulsion, control and communications equipment.[3] Here again the interests, needs, and priorities shaping the information which flows from WBC correspond to some business interests, needs, and priorities of the Westinghouse conglomerate.

Another example from the Westinghouse file relates to other documentary-informational programs produced by WBC. In his 1969 report to Westinghouse stockholders, the corporation's president noted that the company's growth in the 1970s would be enhanced through the addition of new products and services to the Westinghouse portfolio. First on his list was the health care market, a market Westinghouse had just entered in 1969 through the establishment of new subsidiaries. Elsewhere in the 1969 and 1970 annual reports, one learns that of the documentaries produced by WBC in this two-year period, fully one-third were concerned exclusively with the nation's health care system.

Additional examples are evident in the case of the General Tire and Rubber Company. Of the total of seven public affairs programs produced by the company's broadcasting subsidiary during the entire year of 1969, one concerned the matter of safe boating, another the problem of air traffic congestion.[4] Here the very fact that such matters were among only a handful singled out for emphasis raises disturbing questions about General Tire's seriousness in addressing public affairs. At least in the Westinghouse examples the subjects were inherently controversial, whether or not Westinghouse treated them as such, and even if the process of their selection was related to the business priorities of the corporation. On the other hand, one would be hard pressed to find antagonistic views on the subject of safe boating. However, finding clues to this choice of subject is an easier matter. Interestingly, General Tire manufactures most of the vinyl and plastic materials used in pleasure boat interiors, including boat flotation cushions. And an indication of the source of General Tire's interest in air traffic problems is seen in the company's ownership of Frontier Airlines.

Each of these examples, from RCA, CBS, Westinghouse and General Tire, suggests that intraconglomerate business interests shape intraconglomerate information interests; and that these, in turn, ultimately help define the shape and substance of the electronic communications flow. But the incentives manifest here are apparently not the only ones afforded by conglomeration. Other opportunities to coordinate business investments also arise, again with consequences for the shape of the information flow.

Avco, for example, has found a way to launch a new line of services, first by selling the services to its broadcasting subsidiary as program material, eliminating the risk in its diversification investment, and then separately marketing the services to consumers. The service involved is meteorology. It is communicated over Avco stations as an unusually comprehensive brand of weather programming. Avco's stations feature special farm weather programs, sports weather programs, and highway weather programming. Avco acknowledges this exceptional communications specialty by proclaiming in full-page ads that ''Avco Broadcasting is tomorrow's rain, sunshine, cloudy skies, sleet or snow."[5] Such programming has facilitated establishment of a new Avco International Services Division, which began sales in 1970 of technical and management services in radar, meteorology and weather data-handling to private companies. With a mutually promotional twist, the service is marketed as a ''private weather information service provided by Avco Broadcasting's Meteorologists."[6]

It may come as a surprise to learn that the FCC has begun recently to encourage intrabusiness coordination in shaping the communications flow. One example of this policy comes from the Commission's decision in 1971 to approve the merger of the Corinthian Broadcasting Corporation with Dun and Bradstreet, Inc. Corinthian is a group broadcaster owning five television stations, three in the top 50 markets. In hearing the case, the Commission withheld approval until the corporate survivor of the merger, Dun and Bradstreet, came forth with promises of programming changes which the Commission found to be ''significant program improvements."[7] The promises consisted of the corporation's pledge to involve the personnel and products of several of its subsidiaries in new programming series. Dun and Bradstreet's Thomas Y. Crowell Co. and related children's book-publishing interests would provide material for twenty-six programs aimed at preschool children. Another subsidiary, the Life Extension Institute Inc., would provide a series with sixty-five installments programmed around its interests and resources, which center on the geriatrics market. And Dun and Bradstreet further pledged a variety of programming inputs based on the interests of the 18 trade, professional and specialty journals edited and published by its Reuben H. Donnelly subsidiary.[8]

These examples all acknowledge incentives to exploit business opportunities within the conglomerate. Perhaps just as real are incentives to avoid internal conflicts. More specifically, one would not expect broadcasting conglomerates to be openly hospitable to using rare and precious public-affairs time to encourage or fuel controversy about certain of their corporate projects. Some projects, no doubt, can benefit from increased audience awareness. The mass transit and health care documentaries from Westinghouse are suggestive examples. But in other substantive areas, broadcasting subsidiaries are probably well-advised to proceed with caution, lest they jeopardize the smooth functioning of their family of businesses. In a political climate shaped by energy crises, ecological crises, surveillance crises, and weapons race-disarmament crises, the areas of business sensitivity become pervasive.

We, likewise, must also proceed with caution. For one can never be fully certain why the vast number of subjects that are almost totally ignored on the TV screen are ignored; although to be sure, it is more convenient to some business interests and more profitable to others to be shielded from public view.

Since defense contracting by broadcasting conglomerates received the most emphasis in our survey of their diversified businesses, programming related to this subject holds special interest here.

The idea that there exists a military-industrial complex which commands America's domestic economy, foreign policy, and functions generally to prevent peace, is neither a novel nor a particularly recent point of view. Exposition of the idea and opposition to it appeared in numerous books and periodicals well before the Southeast Asian war helped force the term into vernacular.[9] The public may very well equate the term itself with passing political movements such as the Students for a Democratic Society, and movement celebrities like Thomas Hayden. Many people would no doubt be surprised to learn of the active concern about a military-industrial complex registered by such respectable religious organizations as the National Council of Churches and the Society of Friends (Quakers).[10]

But if Hayden's views on the American political process helped make him a suitable subject for media coverage, the same has not been true of the views themselves. Broadcasting conglomerates continue to show conspicuous lack of attention for the whole subject of military-corporate relations and policies.

To say that the networks have failed to consider airing the matter, however, would not be accurate. One innocent CBS documentary producer, Gene De Poris, had already begun research for a public affairs special on the subject when his superiors removed him from the project in 1968. They explained to De Poris that no such complex existed. And indeed, the program that ultimately evolved from the project implied as much. One reviewer described it as a "study of the nomenclature of the ABM."[11]

Another CBS producer who had endeavored to specialize in documentary coverage of military policies was given an indefinite leave of absence by CBS. Peter Davis had achieved broadcasting notoriety in 1971 with his video version of Senator J. William Fulbright's popular book on Pentagon public relations techniques.[12] That hour length documentary "The Selling of the Pentagon," was broadcast by CBS in prime time. The following year Davis produced a documentary on antipersonnel weapons. Subject to drastic cuts by network executives, it eventually was aired as a mini-feature (at the far end of the Sunday afternoon video ghetto) in the CBS news magazine format. Davis left the network later that year to do independent productions on the American military.[13]

One can only speculate what factors prompted all three networks, ABC, CBS, and NBC, to refuse copies of the secret Pentagon Papers, offered to them at the same time the documents were eagerly accepted by newspaper editors.[14] Perhaps former CBS News president Fred Friendly provides a clue: Friendly wrote several years before of the factor he was forced to consider (and which forced his resignation) in comparable situations during his tenure at CBS. He quoted one familiar CBS management position as follows: "there are times when responsible business judgments have to determine how much coverage of the Vietnam War one network and its shareholders can fiscally afford."[15] Friendly has since charged that broadcast television must "share the guilt" for American escalation in Vietnam, not because the escalation went unreported, but as Friendly argued, "we may have been providing most of the parts of the mosaic but in my view we lacked the will and the imagination to relate them to one another."[16]

Under business control, broadcasting has never shown much willingness to provide sufficient public affairs time to either connect Friendly's mosaic or afford a full airing of diverse views on emergent social issues. Thus, to attribute continued avoidance of specific topics to conglomerate control is risky and better left to more bold investigators; or to those who, like Friendly, speak with the authority of experience. And the video screen's silence on matters sensitive to corporate nerves has been sufficiently pervasive to provoke a number of suggestive accounts on this score.[17]

If the logic of business responsibility acknowledged by Friendly is exacerbated under the authority of business conglomerates, are there perhaps checks and balances accompanying conglomeration?

COUNTERVAILING INCENTIVES?

Before diversified conglomeration became imperative to broadcasters, business influence on the substance of communications was regarded as a form of external censorship. During the 1950s, incidents like the deletion of references to gas (chambers) in a gas company-sponsored dramatization of the Nuremburg Trials became legendary.[18] To counter the visibility of the editing and filtering process in innumerable incidents like this, broadcasting executives assumed the posture of wanting maximum professional independence for their producers. In fact, it was this vocabulary which was used by ABC, CBS and NBC in 1960 to justify centralizing under their control the production of all entertainment and information programming they broadcast (see Chapter 5).

But the first concerted push for professionalization in broadcasting originated in the period just before the introduction of television technology. Concerned about the blatant expression of his newscasters' opinions over radio, CBS chairman William Paley settled on a solution. Newscasters would henceforth be labeled analysts, not commentators. The CBS policy meant that newsmen could analyze the news, but promote no views—not explicitly their own at least. Newscaster H. V. Kaltenborn reported how CBS executives explained this policy to him:

> Vice-President Edward Klauber would call me up to his
> office for a friendly heart-to-heart talk . . . "Just don't
> be so personal," he'd say to me. "Use such phrases as
> 'it is said,' 'there are those who believe,' 'the opinion is
> held in well-informed quarters,' 'some experts have come
> to the conclusion. . . .' Why keep on saying 'I think' and
> 'I believe' when you can put over the same idea much
> more persuasively by quoting someone else?"[19]

With most broadcast news during this time focused on America's war efforts, a consensus developed during the 1940s in support of the doctrine of objectivity. The NAB code of ethics called upon broadcasters to report news free of bias. (In the same code, its subscribers were told not to sell air time for the presentation of opinion or controversy.) Three radio networks issued a joint statement declaring their commitment to strict objectivity. And the FCC announced its Mayflower doctrine which said that broadcasting "cannot be used to advocate the causes of the licensee."[20] While the industry denounced the language of Mayflower, it resembled superficially their own pronouncements.

With the passage of time, changes in broadcasting's ownership structure have redefined the source of interests which may shape informational programming. The once tainted influence of sponsors, seen in the Nuremburg example, seems relatively innocuous in the present context of conglomerate incentives. Business interests (business responsibilities, in executive jargon) which were once formally external to broadcasting are now incorporated as communications subsidiaries.

Accordingly, there is a new incentive to emphasize the importance of objectivity in informational communications. Rare is the text on broadcasting or the book-length critique of television which does not profess faith in professionalism as a countervailing influence to business incentives. For years, conglomerate broadcasting executives have used nearly every forum their banquet circuit provides to advance the cause of objectivity and extol the professionalism they claim to be cultivating.

On the one hand, the case for professional incentives in the present business structure is not unlike the FCC's theory that local business ownership provides grassroots responsiveness in broadcasting. Its premise is remarkably unexamined. Broadcast executives promote the creed of objectivity. Most critics accept the creed at face value, only imploring the TV corporations to adhere to an ever more refined sense of professionalism. There are everywhere complaints about the adequacy of its practice, but few inquire into its meaning.[21]

If faith in professionalism is often promoted, one can well appreciate the incentive for promoting that faith. The prevailing business structure in broadcasting makes the defense of informational objectivity compulsory.

Yet there are structural factors which favor the cultivation of objectivity and the development of professionalism. They do not function as countervailing incentives, however, in the sense of checking exclusive business control over broadcasting. They may, rather, enhance it.

Clay T. Whitehead, director of the White House Office of Telecommunications Policy (an agency created in 1970 by President Richard Nixon), observed recently that "those with leadership and decision-making responsibilities must consider information as a major industry, a national resource, and a source of economic and political power."[22] Here Whitehead simply voices the view we began with: that communications inequalities are power inequalities.

However, one is not likely to encounter such candor in those who, by virtue of ownership, have control over the information flow. For to maintain and insure their control, it is advisable that they address the communications system as though it were unrelated to

power. And in a communications system where production and access
are the exclusive rights of one class of interests, be it business or
some other, the propagation of standards of objectivity is one useful
way to moot the issue of exclusive control.

There is an incentive to renew the audience's belief in objective
processing of information, because that belief functions to block audience apprehension of the communication system's operations and its
power. Because it denies the significance of special interests, objectivity legitimizes exclusive control. This is its rationale: as long as
the standard of objectivity is applied to informational processing, it
matters not who does the processing.

The objectivity ethics are routinized through what is called professionalization. Here broadcasting borrows from the postures of
other professions. Professional ethics in communications parallel
the objective neutrality of the scientist in making information processing a skill. As a set of techniques, the science of information processing becomes separated from the person practicing the techniques.
Consequently it would appear that the issue of who controls the information flow is without consequence. Professional ethics have a way
of being self-confirming, since applying the required techniques to
raw information produces roughly the same end result, regardless of
who applies them.

For professionalism in broadcasting, norms are also borrowed
from clinical professions like law and medicine. The norms of de-
tachment and distance help define the communications flow as information prepared by technicians in service to client audiences. As in
clinical professions, role distance contributes to the belief that the
interests of the technician are no different from those of the client
audience.

Through its norms and ethics, professionalism in communications fosters and maintains the illusion that it is unrelated to social
power. Yet communications are inherently manipulative. And exclusive control over communications translates into exclusive opportunities to manipulate. Enzensberger explains this proposition.

> Manipulation—etymologically, handling—means technical
> treatment of a given material with a particular goal in
> mind. When the technical intervention is of immediate so-
> cial relevance, then manipulation is a political act. In the
> case of the media industry that is by definition the case.
>
> Thus every use of the media presupposes manipula-
> tion. The most elementary processes in media produc-
> tion, from the choice of the medium itself to shooting,
> cutting, synchronization, dubbing, right up to distribu-
> tion, are all operations carried out on raw material.

There is no such thing as unmanipulated writing, filming
or broadcasting. The question is therefore not whether
the media are manipulated, but who manipulates them.23

But as he further points out, the

superstition that in political and social questions there is
such a thing as pure, unmanipulated truth, seems to enjoy
remarkable currency. . . . It is the unspoken basis premise of the manipulation thesis.24

Which is to say that those within the audience still able to recognize
opinion in the information flow seem to have so accepted the premise
of objectivity that they insist opinion be made to disappear.
    This characteristic is noticeable in most critics, right and left,
if some of the better known attacks on informational programming
are indeed as representative as they appear.25 If so, it is only further
testimony to how disarming and acceptable the myth of objectivity has
become. Perhaps in awareness of this, some industry technicians
have begun extrapolating this evidently irresistable doctrine to the
general level of public rights in communications. For example, while
federal courts have construed the public's rights in broadcasting to
require "the widest possible dissemination of information from diverse and antagonistic sources,"26 one industry brief to the FCC, the
Litwin study, translated this to mean "the public's right to receive
valid and accurate news and information."27
    Professionalism and objectivity can be subtly important to the
reproduction of exclusive business control and the dominance of its
interests in two further respects. First, the impression is maintained and encouraged that any consequences that flow from broadcasting information are independent of the broadcast medium. This
suggests that broadcasting is somehow not responsible for what it
reports, and what it does not report. The purveyors of this view pre-
tend that the medium is above the historical process. This encourages
the audience to look beyond the medium for clues to the sources of
stability and continuity of established social structures.
    A second effect of the professionally objective facade is that it
obscures consideration of the ways the medium may be instrumental
to the social power of interests other than those presently enjoying
exclusive control. Very simply, it inhibits audience consciousness
of the potential of the resource and its alternative uses.

DIAGNOSIS AND PROGNOSIS

In assessing the cases of intrabusiness coordination, we have purposely spoken about effects on the shape of the information flow. For the architecture of information is quite different when diverse economic classes and social collectivities feed into the information flow, from the information architecture that is created by their exclusion. When labor as well as business feeds into the information flow; when racial, sexual, ethnic, and cultural interest groups feed their informational priorities and needs into the communications system, as well as business—then there emerges a multidimensional, diverse and antagonistic information structure. By comparison, the diversity provided by servicing intrabusiness needs, interests, and priorities is flat. To be sure, the information that flows into the communications system from diversified conglomerates reflects an array of interests and priorities. But it is not a socio-economic, cultural, or political pluralism; it is very much a business pluralism.

If it is a business pluralism, however, it is not quite the one which the FCC envisioned in its regulatory ideals, foreseen in the policies of local ownership and crossownership. It is a transformed business pluralism, redefined in the process of diversification and conglomeration. Moreover, all this suggests that the paradigm for regular paid access to the electronic media, which business owners have designated as the exclusive privilege of business, is being generalized here to include the rest of the communications flow, and most significantly, to include its information component.

It is essential, at this juncture, to understand that a communications architecture defined by business, and attuned to its interests, needs, and priorities, is in no sense a conspiracy. Some defenders of the status quo in communications are all too easily given to dismissing serious critiques of the system as merely conspiratorial theories. But business ownership, exploitation of public communications resources for private profit, group ownership, expansion and diversification, have all occurred in public view, and with the consent and sometimes the encouragement of regulatory authorities. The business incentives which gave cause for each of these developments, and which now function to generalize business access, arise not as an impropriety, but as a routine and legitimate feature of business ownership. These incentives and their exercise are consistent with the logic of business responsibilities and regulatory policies over the course of many decades. Recall that in the case of the Corinthian-Dun and Bradstreet merger, the promise of intraconglomerate business access was viewed as a fortuitous opportunity, one which commanded the general admiration of the Commission.

The fact that it is not a conspiracy, however, is what makes it so dangerous. For as business control over communications is reproduced; as communications control structures become fused with those of the larger industrial system; and as the exercise of the resulting incentives is legitimated by regulatory authorities, we are invariably faced with a communications system which is publicly reckoned to be "normal," and which is, as NAB president Wasilewski has said, "the most successful and universally accepted business enterprise in history." To that degree, the possibility of achieving alternative uses for public communications resources is further foreclosed. And diminishing with it is the potential for a free and open communications flow, of the kind that could emerge from an egalitarian communications structure which respects the rights to free expression of diverse and antagonistic interests, because it is not established to create communications inequalities.

The future of corporate responsibility and conglomerate enlightenment in serving the public is perhaps alluded to in policies pioneered by Avco under this banner.

> Avco Broadcasting has always placed heavy emphasis on community service, but recognizing that social ills today are more acute than they have been in the past, the company has committed itself to assume even greater leadership in this area.
>
> The new role calls for all employees of each station to seek out community problems, and for the station to then take innovative action to help correct them.
>
> As examples of active response to community needs, personnel at Avco Broadcasting radio and television stations accomplished the following last year:
>
> Organized a national panel of disc jockeys to meet at the White House to discuss the misuse of drugs.
>
> Raised and distributed over $300,000 among approximately 100 hospitals to make hospital stays happier for little children.
>
> Raised $50,000 in one day so a home for teenage delinquents could continue to operate.[28]

If this is the logic of business control, corporate inertia on the public's interests may be preferable to such leadership innovations. Here conscientious business practice produces a move to redefine public interest obligations of broadcasters so as to exclude use of the medium itself from the issue of service to the public. If preposterous, this innovation is equally ingenious. For if the public service performance of a communications system can be appraised on the basis of services which have nothing to do with communications, then the pressure for access is deflected and diffused.

Left to the inclinations of the Federal Communications Commission, and to the control of private businesses, our communications resources will not soon, if ever, be used to facilitate a communications democracy. Neither the business pluralism idealized by the FCC early in the history of mass communications, nor the business pluralism subsequently fashioned under conglomerate control, holds promise for ameliorating communications inequalities. Nor will FCC impatience with the existing shape of the communications flow, expressed through new policies which would virtually compel increased informational programming, suddenly mean access for the historically disenfranchised. As the conglomerates are making increasingly clear, broadcasting's business owners, in search of diversity, do not turn to the audience, in a quest to make their medium more of a forum for public interests, needs, and self-expression. Instead they extend access to their family of diverse businesses. In this process, the corporate interest in communications is intensified—in the name of diversity. At the same time, the audience becomes ever more fixed and entrenched, which is to say excluded.

Summoning objectivity and professionalism and intensifying their practice is increasingly seen as the last resort to ameliorate communications inequalities. Yet this is more a palliative than a protective mechanism. Like external forms of censorship ranging from state-controlled censor boards to press councils of the intelligentsia, objectivity and professionalism are programs of internal censorship which function to represent the communications system differently, to redefine it while reproducing it. As attempts to resolve the contradiction of exclusive business control and the inevitability of manipulation, objectivity and professionalism fail. For the contradiction can be resolved neither by censorship nor by masks of mythical objectivity, but only by direct social control. The objectivity-professionalism ideology confronts business manipulation of the media by requiring the manipulators to disappear. It stops short of opening to excluded interests—the audiences—the opportunity for them to join in manipulation.

No doubt the corporate interest in communications will intensify and the exclusion of audiences will continue, with untold refinements, as long as communications frequencies remain the property of business. In a sense then, FCC policies which have historically isolated ownership as the critical determinant in directing the shape and substance of the communications flow prove to be enlightened. But if the importance of ownership is recognized, its significance is lost in policies narrowly conceived to engineer subtle variations within a business ownership structure. Nor is it found in new reforms calculated to stimulate informational programming, in retreat from the ownership question entirely.

Bertolt Brecht, whose astute conceptions of communications equalities and democratic media structures have informed this analysis, also anticipated the deceptive charm of reforms as against the arduous struggle of creating alternatives. More than 40 years ago Brecht wrote

> This is an innovation, a suggestion that seems utopian and that I myself admit to be utopian. When I say that the radio . . . 'could' do so-and-so I am aware that these vast institutions cannot do all they 'could,' and not even all they want.
>
> But it is not our job to renovate ideological institutions on the basis of the existing social order by means of innovations. Instead our innovations must force them to surrender that basis. So: For innovations, against renovation! [29]

At that time, Brecht's vision of a freely accessible communications system was not the only utopia. But in the years since, the utopias of those owning electronic technologies, the Sarnoffs, Goldensons and their corporations, have been actualized. And familiarity brings them normalcy. Meanwhile, the dreams of a Brecht, lacking corporate muscle, have never passed beyond the drawing boards. But they persist as our utopias.

NOTES

1. Broadcasting (May 24, 1971), p. 4.
2. Westinghouse Electric Corp., 1967 Annual Report, p. 7.
3. Data is scattered in Westinghouse annual reports, 1968-71.
4. General Tire and Rubber Co., 1970 Annual Report, p. 21.
5. Broadcasting (November 2, 1970), p. 15.
6. Ibid.
7. Broadcasting (April 19, 1971), p. 43.
8. Ibid.; and Broadcasting (December 7, 1970), p. 23.
9. See bibliographies in: Marc Pilisuk and Thomas Hayden, ''Is There a Military-Industrial Complex Which Prevents Peace?'' Journal of Social Issues 21, no. 3 (July 1965): 70-113; Richard J. Barnet, The Economy of Death (New York: Atheneum, 1969); Sidney Lens, The Military-Industrial Complex (Philadelphia: Pilgrim Press, 1970);

and Seymour Melman, Pentagon Capitalism (New York: McGraw-Hill, 1970).

10. See, for example: National Council of Churches, Church Investments, Technological Warfare and the Military Industrial Complex, A Report Prepared by the Corporate Information Center of the NCC (New York: NCC, 1972); National Action/Research on the Military Industrial Complex, Weapons for Counterinsurgency (Philadelphia: American Friends Service Committee, 1970).

11. Quoted in The Network Project, "Feedback 4: Broadcast Journalism," Performance 1, no. 3 (July/August 1972): 65; see also The Network Project, Control of Information Notebook No. Three (New York: March 1973), p. 34.

12. Senator J. William Fulbright, The Pentagon Propaganda Machine (New York: Liveright, 1970).

13. Marvin Barrett, ed., The Politics of Broadcasting: Alfred I. du Pont - Columbia University Survey of Broadcast Journalism, 1971-1972 (New York: Thomas Y. Crowell, 1973), pp. 8-9.

14. Columbia Journalism Review (March/April 1972): 46-48.

15. Fred W. Friendly, Due to Circumstances Beyond Our Control. . . (New York: Vintage, 1967), p. xxv.

16. Broadcasting (February 1, 1971), p. 9.

17. See Robert Cirino, Don't Blame the People (Los Angeles: Diversity Press, 1971); Nicholas Johnson, How to Talk Back to Your Television Set (Boston: Atlantic-Little, Brown, 1970), pp. 81-96; The Network Project, "Feedback 4 . . .," op. cit., pp. 57-70.

18. Les Brown, Television: The Business Behind the Box (New York: Harcourt, Brace, Jovanovich, 1971), p. 65.

19. Quoted in Erik Barnouw, The Golden Web (New York: Oxford University Press, 1968), p. 136.

20. Ibid., p. 137.

21. A noteworthy study of the day-to-day practice of journalistic professionalism and objectivity is Edward J. Epstein's News from Nowhere (New York: Pantheon, 1973).

22. Quoted by The Network Project, OTP, Notebook No. Four (New York: April 1973), p. i.

23. Hans Magnus Enzensberger, "Constituents of a Theory of the Media," New Left Review 64 (November-December 1970): 19-20.

24. Ibid., p. 17.

25. See, for example: Edith Efron, The News Twisters (Los Angeles: Nash, 1971); and Harry Skornia, Television and the News: A Critical Appraisal (Palo Alto: Pacific, 1968).

26. Associated Press v. U.S., 326 U.S. 1, 20 (1945); see also Red Lion v. FCC, 395 U.S. 367, 390 (1969).

27. George H. Litwin and William H. Wroth, The Effects of Common Ownership on Media Content and Influence (Prepared for the National Association of Broadcasters, July 1969), p. 3.

28. Avco, Inc., 1969 Annual Report, p. 26.

29. Bertolt Brecht, Brecht on Theatre, John Willett ed. and trans. (New York: Hill and Wang, 1964), pp. 52-53.

RICHARD BUNCE is a Research Sociologist with the Social Research group, School of Public Health, University of California, Berkeley.

His personal encounters with mass communications have involved contrasting roles with several organizations. He has served as Research Projects Director with the National Citizens Committee for Broadcasting. He has also worked with a community videotape production group, with the editorial collective of an irregular journal of writings on men's liberation, and, on occasion, has appeared with community groups as an on-the-air advocate of particular issues.

He did his graduate work in sociology and law at the University of Wisconsin, Madison.

CITIZENS' GROUPS AND BROADCASTING
Donald L. Guimary

COMMUNICATIONS AND PUBLIC OPINION: A
Public Opinion Quarterly Reader
Robert O. Carlson

EDUCATIONAL TELEVISION FOR DEVELOPING
COUNTRIES: A Policy Critique and Guide
Robert F. Arnove

THE ELECTRONIC BOX OFFICE: Humanities
and Arts on the Cable
Richard Adler
Walter S. Baer

FREEDOM OF THE PRESS VERSUS PUBLIC ACCESS
Benno C. Schmidt, Jr.

FOREIGN AFFAIRS NEWS AND THE BROADCAST
JOURNALIST
Robert M. Batscha

INTERNATIONAL COMMERCIAL SATELLITE
COMMUNICATIONS: Economic and Political Issues
of the First Ten Years of INTELSAT
Marcellus S. Snow

THE MEDIA AND THE LAW
edited by Howard Simons
Joseph A. Califano, Jr.

PUBLIC ACCESS CABLE TELEVISION IN THE
UNITED STATES AND CANADA: With an Annotated
Bibliography
Gilbert Gillespie

TELEVISION AS A SOCIAL FORCE: New
Approaches to TV Criticism
Richard Adler